MEANINGS AT THE TEXT LEVEL: A CO-EVOLUTIONARY APPROACH

CONTEMPORARY STUDIES
IN DESCRIPTIVE LINGUISTICS

VOL. 2

Edited by

DR GRAEME DAVIS & KARL A. BERNHARDT

PETER LANG
Oxford • Bern • Berlin • Bruxelles • Frankfurt am Main • New York • Wien

T. Nyan

MEANINGS AT THE TEXT LEVEL: A CO-EVOLUTIONARY APPROACH

PETER LANG
Oxford • Bern • Berlin • Bruxelles • Frankfurt am Main • New York • Wien

Bibliographic information published by Die Deutsche Bibliothek
Die Deutsche Bibliothek lists this publication in the Deutsche Nationalbibliografie; detailed bibliographic data is available on the Internet at ‹http://dnb.ddb.de›.

British Library and Library of Congress Cataloguing-in-Publication Data:
A catalogue record for this book is available from The British Library, Great Britain, and from The Library of Congress, USA.

ISSN 1660-9301
ISBN 3-03910-250-8
US-ISBN 0-8204-7179-0

Hochfeldstrasse 32, Postfach 746, CH-3000 Bern 9, Switzerland
info@peterlang.com, www.peterlang.com, www.peterlang.net

Printed in Germany

For John

Contents

Preface 9
Acknowledgements 11

Introduction 13

Deacon's co-evolutionary approach 15
Viewing meaning from a co-evolutionary perspective 20
Text level, communicative situations, and language evolution 20
Problems and underlying means 26

Chapter 1

Illocutionary acts and divergent thinking 29
Direct illocutionary acts with a declarative structure 36

Chapter 2

SN conditions and the internalization of environmental constraints 59

Chapter 3

Argumentational meaning 89
The knowledge source problem and decision making 95
The interpretative problem: argumentation, categorization and divergent thinking 112

Chapter 4

Methodological issues 155
Edelman's theory of neuronal group selection 166
Meanings from the perspective of the TNGS 176

Bibliography 179
Index 185

Preface

From a linguistic standpoint, speech acts, SN conditions (conditions for the success and non-defective performance of speech acts) and argumentational meaning present too few commonalities for their emergence to be amenable to a unified account.

Such an account, nonetheless, would appear to be possible, if, following Deacon (1997), one assumes that language and the brain are co-evolved, and that language, as a latecomer, can draw on a pre-existing stock of solutions when faced with problems of its own.

The above meanings – I hypothesize – arise as the end-result of language adaptation to processing efficiency, a requirement to which all processing systems are subject.

In terms of what subserves this emergence, a co-evolutionary approach allows for the possibility that pre-existing strategies, processes and constraints arising from brain systems have been co-opted, which have proved successful in dealing with similar problems.

Such an approach thus provides some clear indications as to how to get started: identify the problem and this will point you in the direction of likely underlying means.

In connection with the above meanings, four types of processing problem, specific to the text level, can be singled out:

(1) How to hit upon an interpretative short-cut between lower-level structures (associated with lower-level meanings) and higher-level meanings?
(2) How to retain the lower-level meaning and integrate it to the higher-level meaning?
(3) How to gain quick access to relevant contextual cues?
(4) How to go about identifying the upcoming text?

In response to these problems, I propose that likely candidates include divergent thinking, various processes inherent in the perceptual systems, whereby environmental constraints are internalized, and processes involved in neurobiologically rooted models of decision-making.

This type of account – it goes without saying – must rest on known properties of brain structure. A further condition, however, is that the global brain theory (or family of theories), which provides them with their characterization, should be consistent with a co-evolutionary perspective. This has the consequence of ruling out an information processing or functionalist model of brain function (i.e. one defined in terms of the internal functional organization, or pattern of cause-and effect relations) in favour of a model that is biologically based, in the sense that it relates function (e.g. categorization) to physical features (e.g. neural structure).

Acknowledgements

This study would not have been possible without the generous support I received from the Rockefeller Foundation.

I would also like to express my thanks to the many people who have helped me in various ways and at various stages. Among those I am especially grateful to Daron Burrows, Bill Croft, Marina Colasanti, Robert MacLaury, Anne Morton, Maj-Britt Mosegaard Hansen, John Searle and Helena Sedgwick. Finally, my deepest thanks go to Marilyn, Harvey and David Schwartz for their unwavering support and exceptional hospitality.

Introduction

Certain linguistic[1] meanings arising at the text level are not predictable on the basis of constituent structure alone. These include speech acts, conditions for the success and non-defective performance of speech acts (SN conditions), and certain types of constraint on continuation, known as 'argumentational meaning'. Consider the following examples:

(1) Prunella is a mezzo-soprano.
(2) I will get you a season ticket.
(3) *You will shut the door on leaving, but I know you can't do it.
(4) *I will make sure it gets to you in time, but I have no intention to do so.

1 Building on Levinson (1995), Traugott (1999, pp.94–5) proposes the following categories: (a) coded meaning (semantics); (b) utterance-token meaning; (c) utterance-type meaning. The first category results from conventions of language. The second consists of invited inferences (IINs), which are approximately equivalent to conversational implicatures and defeasible; the third corresponds to generalized invited inferences (GIINs). Approximately equivalent to generalized conversational implicatures, GIINs are preferred meanings arising from conventions of use, which are defeasible. 'Linguistic meaning', as used in this study, not only includes coded (semantic) meaning, but also preferred 'pragmatic' interpretations arising from use conventions. The rationale behind this is that from a co-evolutionary perspective, there can be 'partial' coding, involving categories of varying degrees of abstraction, and mere shifts to a different domain of reference, alongside precise contents. Furthermore, where traditionally it is assumed that the 'site of encoding' is a syntactic representation, I will propose that it can also be a 'motor representation'.

(5) a Pierre est riche.
Pierre is wealthy.
b Il peut s'offrir n'importe quoi.
He can afford anything he fancies.
c Il est donc avare.
He is therefore stingy.

(1) and (2) are linguistic strings conventionally associated with a propositional meaning (purporting to represent a present and future state of affairs, respectively). The utterance of the former expresses an assertion, that of the latter, an act of promising. The sequences in (3) and (4), which are self-contradictory, are meant to illustrate the fact that some SN conditions (in the sense of Searle 1969 and 1983) form an integral part of the logical structure of speech acts. What is being denied is a preparatory condition in (3) and a sincerity condition in (4). (5), borrowed from Anscombre (1995), is a typical example of argumentational meaning, a type of weak constraint on continuation.[2] In virtue of this constraint, (b), but not (c), constitutes a preferred continuation to (a), in a context free situation.

How did this come about? How do strings associated with propositional meaning (arising from their constituent elements) come to be encoded with the above meanings at the text level (where 'text' includes spoken and written propositional discourse)? The fact that we are dealing with varying degrees of encoding can be ignored for the time being.

As stated, the problem could be construed as the province of Historical Linguistics. Current research in meaning change, according to Traugott (1999, pp.92–3) 'seeks to identify regularities in paths of meaning change, and the motivations that lead to these changes'. Central to this research there appear to be three questions:

2 In this particular instance, the constraint arises from the use of a lexical item ('riche'), but earlier sources for it include argumentative markers such as *mais* (but), *même* (even), *d'ailleurs* (besides).

(1) Given the form-meaning pair L (lexeme), what changes did meaning M undergo?
(2) Given the concept C, what lexeme can it be expressed by?
(3) Given the concept C, what other conceptual domain is it likely to develop into?

From a historical standpoint, these questions would be susceptible to an answer in terms of the nature of the derivational processes, paths of evolution, conceptual categories, direction of change, shift in conceptual domains, motivating communicative functions.

By contrast, the questions I seek to address concern underlying capacities, strategies, processes, and constraints arising from brain systems, which, on a co-evolutionary approach, underlie the emergence of the above types of meaning. Thus, with respect to speech acts, SN conditions and argumentational meaning, the basic question is: what pre-existing means make it possible for strings expressing a lower-level meaning to become further encoded with certain higher-level meanings, including various types of contextual meanings? This type of question can only be posed within a framework such as Deacon's (1997), which takes as its basic assumptions that:

(a) language and the brain are co-evolved;
(b) language, as a latecomer, availed itself of whatever pre-existing means were best suited to its needs.

Deacon's co-evolutionary approach

Rather than provide a comprehensive presentation of Deacon's theory, I shall confine myself to issues having a direct bearing on our purposes, in particular that of innateness. In stark contrast with Chomsky, Deacon places innateness at the core of a full-scale theory of evolution. In Chomsky it features as a mere assumption – albeit a foundational one, leading to a new paradigm in linguistics. Chomsky, famously, sets out to describe language competence. Deacon, on the other hand (1997, p.35), does not regard children's grammatical abilities as 'the crucial mystery of language'. More fundamental, in his view, are:

(a) the absence of simple languages used among other species, which appear to be capable of many 'other equally or more complicated modes of communication';
(b) the fact that it seems virtually impossible for them to learn a simple language, in the sense of 'a mode of communication based upon symbolic reference (the way words refer to things) and involving combinatorial rules that comprise a system of representing synthetic logical relationships among these symbols.'

To return to the question of innateness, what is at issue is not that human brains come into the world with predispositions to learn language. Rather, it is that those predispositions should be the result of an innate language competence, a grammatical knowledge prior to experience. Deacon objects to the hypothesis of an innate Universal Grammar on grounds that it makes the wrong assumptions about brains and evolution and inverts cause and effect. Where Chomsky and his followers typically 'assert that the source of prior support for language acquisition must originate from inside the brain', Deacon (1997, pp.105–15) offers this alternative, that 'the extra support for language learning is vested neither in the brain of the child nor in the brains of parents or teachers, but outside brains, in language itself', whose evolution is considerably more rapid than brain evolution. 'Children's minds need not innately embody language structures, if languages embody the predispositions of children's minds.' The structure of a language would be to the brains of children what a virus is to its human host, in that it too is under intense pressure to adapt to its human hosts, on which it depends for its reproduction. Pursuing this analogy further, Deacon goes on to suggests that, similarly to what happens in the case of the virus and its human host, the relationship between language and the brain, will have evolved into a form of symbiosis, for the mutual benefit of both parties:

> we should not be surprised to find complex human adaptations to language on one hand, whose purpose is to ensure that language is successfully replicated and passed from host to host, and language adaptations to children on the other, whose purpose is to make languages particularly 'infective' as early as possible in human development. Modern humans need the language parasite in order to flourish and reproduce, just as it needs humans to reproduce (Deacon 1997, p.113).

Greater intelligence, articulatory abilities, and grammatical predispositions in children were not, in his view, prerequisites of language evolution, but evolved as supports for this novel form of communication. Once the first system of symbols was created and found indispensable, brain evolution would have followed in response to this selection pressure. This in turn would have allowed for language to evolve to its modern level of complexity. In short, the ensuing picture is one of co-evolution of language and the brain.

Language change also plays a key role in Deacon's position with respect to language universals, the existence of which once provided another powerful argument in support of theories of innate linguistic knowledge. Given the indirectness of the link between psychological universals and linguistic universals, the latter need not be hard-wired into the brain, but can more plausibly have arisen from a wide variety of factors, of which innate mental tendencies are but one. Language universals 'have emerged spontaneously and independently from each evolving language, in response to universal biases in the selection processes affecting language transmission', such as 'immature learning biases, human mnemonic and perceptual biases, the constraints of human vocal articulation and hearing, and the requirements of symbolic reference' (Deacon 1997, p.116). Language universals are what remains when the 'less learnable variants' have been selected against because they are less adapted to human minds, that is, to pragmatic constraints governing language uses as well as to the human nervous system:

> No matter which formal, cognitive, or phonological universals of language we consider, they probably arose as a result of biases affecting nongenetic evolutionary processes of language in a manner that is similar to the way that the designation of colors in different languages tends to converge on the same referents and multiply color terms along parallel lines (Deacon 1997, p.121).

In summary, within this co-evolutionary perspective, languages do most of the adapting, and language change is a key factor in understanding the ease with which children acquire a first language and the existence of language universals. If a full evolutionary account is what one is after – which, does not appear to be the case with Chomsky – the postulation of innate linguistic knowledge and a formal description of competence amount to giving a label to and describing what is missing. What a language evolution theory needs to explain must include a scenario of how selection might have favoured the evolution of an innate linguistic knowledge or an alternative for it:

> It must also provide a functional account of why this particular organization was favored, how incremental and partial versions were also functional, and how structures present in nonhuman brains were modified to provide this ability (Deacon, 1997, p.38).

This last point takes us back to what Deacon regards as a more fundamental mystery than children's linguistic learning abilities, namely the fact that learning even a simple language presents an almost insurmountable obstacle for other species. Why this should be so, Deacon puts down to two types of difference: one between human and non-human modes of communication, the other, between human and non-human minds. As we have already seen, the difference between human and non-human modes of communication is a matter of symbolic versus non-symbolic modes of reference; what separates human and non-human minds is that the former but not the latter possess the ability to learn symbolic associations. Basically 'the ability to overcome the symbol-learning problem can be traced to the expansion of the prefrontal cortical region and the preeminence of its projections in competition for synapses throughout the brain.' This apparently resulted from a shift in connectional patterns, due to a 'divergence in size of the human dorsal versus ventral forebrain, […] in ways that are radically different from other species' brains'. This shift is also crucial to understanding another central feature of the human language adaptation, namely the ability to 'produce skilled vocalizations', which 'can be traced to changes in motor projections to the midbrain and brain stem' (Deacon 1997, p.220). That the cortical regions should have been recruited for the language function can be understood on the basis that 'what they were already doing offered the best fit to the new problem posed by language' (Deacon 1997, p.287).

The significance of this prefrontal[3] dominance is that it leads to a predisposition in our species to favour one learning strategy over competing tendencies, whenever the opportunity presents itself. Underlying this cognitive tool is the capacity for 'divergent thinking', whose impairment, in case of prefrontal damage, results in an inability on the part of patients to perform highly specific types of task.

> Tasks sensitive to prefrontal damage [...] involve short-term memory, attention, suppression of responses, and context sensitivity. But they all have one other important feature in common: each involves a kind of negation relationship between stimuli or stimulus-behavior relationships. They all have to do with using information about something you have just done or seen against itself, so to speak, to inhibit the tendency to follow up that correlation and instead shift attention and direct action to alternative associations.

Prefrontally damaged patients appear to fail at tasks requiring them to change criteria (e.g. card sorting), or not to produce again alternatives previously generated (e.g., providing word lists). They also 'have difficulty learning mazes based on success/failure feedback, making plans, spontaneously organizing behavior sequences, and performing tasks that require taking another's perspective'. These various tasks involve thinking in allocentric terms, that is using 'egocentric information [...] as a frame of reference', while performing 'a systematic mental reversal of response tendencies', which requires inhibiting responses based on egocentric information in order to generate alternatives that result from an inversion of those responses (Deacon 1997, pp.263–4).

Divergent thinking also underlies the shift from an indexical mode of reference to a symbolic one:

> The shift from associative prediction to symbolic prediction is initially a change in mnemonic strategy, a recoding. It is a way of offloading redundant details from working memory by recognizing a higher-order regularity in the mess of associations, a trick that can be accomplished without having to hold all the

3 See also Joaquim Fuster's *The prefrontal cortex: anatomy, physiology and neurospsychology of the frontal lobe* (New York, 1980).

> details in mind. [...] The crucial point is that when [...] a systematic set of tokens becomes available, it allows a shift in mnemonic strategy that results in radical transformation in the mode of representation. What one knows in one way gets recoded in another way. It gets re-represented. We know the same associations, but we know them also in a different way (Deacon, 1997, p.89).

Viewing meaning from a co-evolutionary perspective

Without the meanings under study, one can safely assume that language processing at the text level would have been too slow to meet the needs of communication adequately. For this reason, their emergence can be taken to be a response to the requirement of processing efficiency, to which all processing systems are subject. An optimal response to this requirement lies in automatization, a process whereby a direct link is established between input and output points, as exemplified by stimulus-response connections. Where the implementation of this model proves to be difficult – as is arguably the case at the text level – alternatives have to be found to provide shortcuts between intermediate input and output points. How does one set out to find suitable candidates? By first identifying what processing problems arise at the text level.

Text level, communicative situations, and language evolution

As a preliminary to identifying these problems, it would be helpful to have a general idea of the communication situation at the text level, especially of how it came to be what it is. 'Text level' stands for those levels of structural complexity which fall under what Givón (1979, p.223) calls the 'syntactic mode' of communication. This mode, which is opposed to the 'pragmatic mode', ontogenetically and phylogenetically earlier to it, apparently arose out of 'four major coding modalities, rigid word-order, grammatical morphology, constraints' (in the sense of conditions such as 'coreference, case-identity, parallel structure'), and 'intonation' (Givón 1979, pp.232–3). Of these four modalities, the first three are the results of the process of syntacticization, whereby 'loose, paratactic, "pragmatic" discourse structure develop – over time – into tight, "grammaticalized" syntactic structures' (Givón 1979, p.208). The types of meaning under consideration would have emerged as part of a tendency towards 'further symbolic recoding' at the text level. Such a recoding may be seen as a form of lexical coding, one which presupposes syntacticization. Indeed, it would have arisen in response to a more developed communicative situation than the one Givón describes for syntacticization, in the sense that it contains – among other things – more information, and involves a more complex communicative goal.

How did this new development come about? In order to have a clearer picture, we need to view this 'post-syntacticization' communicative situation in the context of changes undergone by communicative situations, and their correlates in terms of language evolution. According to Givón (1979, pp.299–301), interlocutors, at the earlier stage, would have shared both 'generic knowledge' (i.e. stereotypical knowledge about 'who is likely to do what to whom under most circumstances'), and 'specific knowledge' about the immediate speech situation. To the extent that this background knowledge was easily accessible, it did not require verbal coding. Furthermore, communicative contents would have been confined to what was present on the scene, which means that only the least predictable part of the message, the assertion, was coded (the topic, at that stage, still went uncoded). Gradually, as 'communication among intimates' gave way to 'communication among strangers', there arose a need for background knowledge to be coded verbally, hence the emergence of multipropositional discourse, and with it a different sense of context, namely 'previous context'. 'Our multipropositional discourse is thus the communicative vehicle carrying these complex series of temporal orderings of event and state propositions.' At that stage, the range of communicative contents had also broadened to include 'nonpresent (both in time and space) third persons as topics of the communication, not only as objects-goals but also as agents', hence the need for further coding. To this might be added meta-linguistic information about the relationships between propositions. Why this should be required can be understood on the basis of new demands being made on the processing system, as a result of the higher level of coding brought about by an expanding domain of reference.

In light of what precedes, two points need to be made that are crucial to one's understanding of how the next stage would have come about. As the domain of reference expands to include non-immediate topics, the possibility arises for the emergence of a 'subjective' domain of reference: things in the future can hardly be said to be part of the external world; to exist at all, they would have to feature as intentional objects of mental states such as desire or intention. In other words, inherent in this expansion of the initial domain, we have a subjective angle of viewing susceptible of giving rise to a subjective domain of reference, should the focus of attention shift to relationships between future events and events that would be present on the scene. In short, what is crucial about the emergence of a new domain is that it can be linked with a shift in point of view, or, alternatively with viewing the same things under a different aspect. The other point (about which we will have more to say in due course) is that this different aspect would have to belong to a higher level of abstraction of a certain kind for there to be a further symbolic threshold.

To return to the evolution of the communicative context, the next stage is what we have referred to as the post-syntacticization communicative situation. This stage is marked by an attention shift to relationships between future and present events, such as future actions and present mental states (e.g. catching a plane tomorrow and the desire or intention to do so), or future actions and events co-temporal with the utterance act (i.e. speech acts), which are responsible for bringing these actions and events about. An example is the relationship holding between catching a plane tomorrow and the promise to do so, as performed in saying 'I will catch a plane tomorrow' (which expresses the intention to do so). In other words, the communicative contents characteristic of this stage would consist of the intricate network of relationships involving speech acts, interlocutors, as well as future actions on the part of the latter.

With the causal aspect of these future actions coming into focus, one sees the emergence of two new domains of reference: (a) the speech situation, as a subdomain of the initial domain of reference. (b) a subjective domain. The former, an expansion of the initial domain, comes about as further entities from the external world – speech participants and speech acts – become part of the communicative contents; the latter, as a result of further inclusions into the communicative contents – this time, of relationships between the above entities, in particular, of a causal kind.

Thus, associated with speech acts are two simultaneous changes affecting domains of reference: on the one hand, an expansion of the original domain to include the speech situation as part of the external world; on the other, the emergence of a subjective domain brought about by the introduction of a subjective angle of viewing. Both changes would have come about in response to a more manipulative form of communication, one requiring more emphasis on motivations for expected future actions. Performative sentences (e.g. 'I promise to come') trade on this double status of speaker and act of utterance as elements of both domains.

The postulation of a subjective domain, where relationships between entities are profiled, is consistent with the intuition that at the multipropositional level the basic unit of meaning is no longer propositional, but should be construed as a reference point-content pair, where the reference point is normally a speech participant committed to varying degrees to the content, as viewed under some specific aspect. Furthermore, the introduction of a subjective angle of viewing helps clarify two crucial aspects of discourse: (a) why, at this level, propositional meaning recedes to the background, and speech acts and interrelational aspects of meaning, in general, move to centre stage; (b) the fact that cohesion, as a dynamic semantic relationship between contents, appears to require a reference point, in the sense that it can only arise between contents that are perceived, as it were, from the same angle of viewing.

Now, what about background knowledge? With communicative goals becoming distinctly more manipulative, the problem of ensuring processing efficiency can no longer be posed simply in terms of accessibility: how to get H to use a common evaluation context becomes more crucial. This is where the establishment of direct links, as characteristic of automatization, is called for, (a) to provide direct access to idealized contexts; (b) to constrain H's responses. The presence of in-built SN conditions and of constraints on continuation, in the conceptual structure of certain speech acts, would be the result of such links. A further example can be found in certain argumentational phenomena. Why, in addition to (a), (b) should be needed is because in a situation where communication is action-oriented, providing automatic access to an idealized context may fall short of ensuring that the conditions of satisfaction of one's speech act will be fulfilled. If S wants H to engage in a future action A, providing him with automatic access to an idealized context in which A is justified will no doubt take him some way in the right direction; however, inducing in him a mental representation of that context is no guarantee that he will automatically adopt it as his own, or as a true representation of the situation which is manifest to him. What is required is a higher degree of efficiency, hence a stepping up of the automatization process to include H's responses. Constraints on response and continuation would have arisen to cater for this need. Some of the argumentational phenomena we will be considering could be construed as the result of this strategy being applied to actual continuations.

Given that an expansion of the communicative contents will normally call for further metalinguistic coding to sustain an appropriate level of processing efficiency, a parallel development of the metalinguistic domain can be expected, when the initial domain or a subdomain of it – in the event the speech situation – reaches a critical stage; one such stage comes about with the need to process multipropositional discourse. The way the metalinguistic domain would have evolved to cater for this situation is in the form of discourse markers, also referred to as 'conjunctions' (cf. Robin Lakoff 1971, Sweetser 1988), 'connectors' (cf. Ducrot 1983, Blakemore 1987), 'metalinguistic operators' (Nyan 1998), or 'discourse particles' (Fischer, to appear) depending on which levels of analysis are being privileged. These markers thus appear to constrain the relations between propositions, the nature of the elements involved in a conjunction (e.g. propositional contents, speech acts, assumptions of various sorts), contexts of relevance, as defined by stereotypical contextual assumptions, prior schematized situations or inferential patterns, as well as sources of information required to instantiate them.

To summarize, building on what Givón (1979) proposed in the way of communicative pressures responsible for the rise of syntax, I have suggested a further type of situation, continuous with the previous one, but different from it in terms of communicative contents and domains of reference. As far as communicative contents are concerned, the range of topics came to be extended to future events and to their link to those that are co-temporal with the act of utterance. As a result of this extension, one sees the emergence of two new domains of reference: one, the speech situation, which is really a subdomain of the initial domain, and the other, which arises from the subjective angle of viewing inherent in the new communicative contents. In parallel to these we have a further development of the metalinguistic domain, which, though peripheral from the standpoint of S's communicative intention, is essential as far as the processing system is concerned. As communicative contents increased and became more complex, there appear to be two alternative modes catering for the same type of efficiency, as determined by the manipulative level of communication: one, grammatical, exemplified by syntactic means in general and functional words in particular; the other, lexical, involving a process whereby external elements become part of the conceptual structure. These two modes stand in stark contrast, as they operate at opposite ends of the communicative process, with the former constraining interpretation, and the latter H's responses; furthermore, the former tends to be instructional, in the sense of conveying metalinguistic information about linguistic entities and processes, schematized situations, access order in respect of preferred interpretations and sources of information required for instantiation (cf. Nyan 1993, 1999), while the latter is in the form of constraints, and gives automatic access to certain types of interpretation and response, with a high degree of instantiation.

Problems and underlying means

In light of what precedes, a basic problem facing the interpretative process at the text level is how rapidly to reach a higher level interpretation on the basis of a linguistic structure conventionally associated with a propositional meaning. Thus, in view of the fact that the string 'Prunella is a mezzo soprano' is conventionally associated with the interpretation that it constitutes a representation of a state of affairs, the basic problem is one of setting this interpretation aside in favour of a higher-level interpretation (e.g. to the effect that S's intention in uttering the string at hand is to make an assertion). To complicate matters further, the lower-level meaning is not to be ruled out but retained as part of the higher-level meaning, in the sense that S's assertion is about the representation in question matching an actual state of affairs. In connection with the lower-level meaning, another problem is that of deciding which aspect is relevant to the higher level; thus, if 'Prunella is a mezzo soprano' were to be used as an argument, its representational aspect cannot be the one relevant to its interface with the higher-level meaning. In addition, there is the further complication of bringing to bear appropriate contextual information in order to select the correct speech act, or the correct up-coming co-text, as the case may be. In other words, processing efficiency at the text level would be seriously compromised, unless the interpretative process can, as a matter of routine:

(1) set aside propositional meaning;
(2) shift attention from one aspect of lower-level meaning to another;
(3) shift attention to higher-level meanings;
(4) have easy access to relevant contextual information (ranging from SN conditions to background knowledge of a more general kind);
(5) reach a quick decision when faced with a range of options.

In more general terms, this translates into a need for:

(a) a strategy, whose function is to provide a direct link between lower-order associations of one kind, and higher-order associations of another;
(b) a mnemonic device, whereby lower-order associations can be kept from interfering with higher-order processes;
(c) a mechanism, whereby attention can be redirected from one aspect of a given object, to another;
(d) a process, whereby elements pertaining to the context of occurrence of a stimulus can become associated with the properties of that stimulus;
(e) a source of contextual knowledge;
(f) quick access to this source;
(g) a decision-making process.

Of these, (a), (b) and (c) point to divergent thinking as a likely candidate for co-option; (d), to a strategy similar to one known as 'internalization', whereby environmental factors become internalized by organisms, as part of a process that replaces flexible adaptive responses to such factors by genetic predispositions; (e) suggests the long-term memory system; (f), a process, whereby shortcuts can become established, via the strengthening of synaptic connections. Finally, a suitable candidate for (g) would be a model of decision-making already subserving the requirements of action. It goes without saying that the above means call for further specification, in terms of supporting processes and structures. These will be provided as we go along.

In terms of organization, Chapter 1 is devoted to speech acts, Chapter 2 to SN conditions, and Chapter 3 to argumentational meaning. In each case our concern will be with underlying means. The last chapter raises issues of methodology and evaluation.

Chapter 1

Illocutionary acts and divergent thinking

This chapter is devoted to our first case of language adaptation to processing requirements. Our hypothesis is that divergent thinking underlies the discovery of the shortcut that makes it possible for structures conventionally associated with propositional meaning – or rather their production – to become directly linked with a higher-level meaning, in the event, an illocutionary value.

Preliminary remarks

As the capacity underpinning the switch from indexical to symbolic reference, divergent thinking – as we recall – enables people to inhibit a tendency to respond to immediate stimuli and to direct their attention to a sampling of higher-order alternatives. In the case of symbolic recoding, which was prompted by a need for a more efficient way of representing information, this inhibitory process would be in part responsible for a shift from lower to higher-order relationships as objects of coding. Thus, with regard to the first symbolic recoding, the shift would have been from relationship between words and entities in the world, to that between words – or rather the concepts they express – with shared semantic traits providing suitable contents for recoding. Consider the case of *apple* and *seed* (adapted from Deacon 1997, pp.69–101). Corresponding to the former type of relationship we have the correlative link between the word *apple* and the class of its potential referents; an example of the latter is found in the associative link between the two words, which share the semantic trait, */edible/*. In this scenario the word *food* would have resulted from the recoding of the trait in question. Once the word *food* had become available, direct reference could be made to what hitherto could only be referred to indirectly by means of *apple* and *seed*. Inasmuch as the processing problem encountered at the text level bears a strong resemblance to that leading to the first symbolic recoding, it is not unreasonable to hypothesize that divergent thinking would again have been co-opted to resolve it, thereby giving rise to a further symbolic recoding. Underlying this hypothesis are two fairly standard assumptions, namely:

(a) From an evolutionary standpoint, solutions that have proved useful in dealing with certain types of problem are likely to be called upon again, should the same types of problem present themselves.
(b) There is a tendency on the part of our species to give preference to divergent thinking whenever the opportunity arises.

This same tendency leads one to expect that, synchronically, divergent thinking would also be found at work, as appears to be the case, if the interpretation of speech acts is something to go by. Thus, to arrive at the interpretation that the intended meaning of a declarative utterance is to mark a speech act, one must be able to set aside its propositional meaning and shift attention to the meaning produced *in* saying the utterance in question. This process, whereby the focus of attention is redirected from a lower-level to a higher-level meaning, is also exemplified by the interpretation of indirect speech acts. Consider:

(1) Can you pass the chili?

In virtue of a certain convention of use, questions such as (1) are often taken to be requests. However, for this indirect illocutionary value to emerge, the lower-level value – a request for information about H's ability to do something, in the event, to pass the chili – must be set aside. A similar shift is required in the interpretation of conversational implicatures, for example, in deriving /Do something about it!/ from:

Ludwig needs a debriefing.

Criteria for divergent thinking

Inasmuch as divergent thinking involves an attentional shift from lower to higher level, a general requirement is that possible candidates for lower and higher levels (i.e. objects of perceptions susceptible of being so construed) should belong to a coherent conceptual domain.[1] This preliminary requirement, which is about observability, entails that the criteria we now turn to are only applicable to cases that synchronically lend themselves to this kind of construal.

Criterion 1

Lower and higher levels should stand in a hierarchical relationship, as defined by the domain of application. Indeed, divergent thinking typically underlies tasks involving subordinating elements to one another (e.g. symbol learning, which crucially depends on an ability to subordinate one associative learning process to another (cf. Deacon 1997, p.264)).

Criterion 2

When viewed in terms of the higher level, the lower level should be seen under a different aspect to the original: as will be recalled, the purpose of the attentional shift is to allow a sampling of allocentric, rather than egocentric alternatives, which entails a change of perspective.

Criterion 3

The directionality of the attentional shift should be from a domain of reference that is closer to (or identical with) the external world, to one

1 This conceptual domain need not correspond to dimensions of experience, as normally construed, but could be theory-bound.

further away from it.[2] This is to do with the nature of the communicative contents and communicative goals at the text level: since the communicative contents expand to include entities not present on the scene as well as relationships, and the communicative goal becomes interactive, it is hardly surprising that the things being referred to should pertain to domains of increasing levels of abstraction.

Criterion 4

There should be an identifiable problem for which shifting attention to the higher level is conducive to a better solution. Our assumption is that divergent thinking, being goal-oriented rather than goal-seeking, is concerned with means, whereby a pre-set goal can be achieved.

The above criteria, the satisfaction of which leads to a situation favourable to a co-opting of divergent thinking, are susceptible to further specifications arising from particular domains of application. The one under consideration is that of meanings at the text level, with propositional meaning and speech acts providing possible reference points for lower and higher levels.

Concerning Criterion 1, the hierarchical relationship finds an obvious instantiation in the logical structure of intentionality (cf. Searle 1973, pp.4–26). Speech acts, as a form of intentionality, share with intentional states a structure consisting of a 'psychological mode' and an 'intentional or representative content', except that in referring to them it is more customary to talk about 'illocutionary force' and 'propositional content'. Thus, just as the belief that John has a new

2 Regarding what constitutes an abstract domain, Croft (2000, p.260), following Langacker 1987, has this to say: 'abstract domains are [themselves] components of a very large number of concrete domains. For example, almost all domains make some reference to time or scales in defining the concepts profiled in them (any domain involving events and gradable properties respectively), and an enormous number involves some reference to spatial concepts in their structure (any domain involving physical objects or motion). It is more precise to think of domains as being themselves multidimensional […] the "dimensions" that are found in the largest number of domains are the most abstract ones.'

dog would be analyzed in terms of a psychological mode, belief, and an intentional content to the effect that John has a new dog, so would the assertion 'John has a new dog' be analyzed in terms of an illocutionary force, assertion, and a propositional content, /John has a new dog/. Another aspect of speech acts that is relevant to our purpose is that each type is the expression of a mental state, which is its sincerity condition, and – more to the point – both speech acts and mental states share the same intentional content, which expresses their conditions of satisfaction.[3] Thus, if I promise[4] to come, my act of promising expresses an intention on my part to come, which is the act's sincerity condition, and both the act and the intentional state are satisfied if I come. Similarly, the order that H should get out is the expression of its sincerity condition, namely, S's desire that H should get out, and H getting out would satisfy the conditions of satisfaction of both order and desire. Finally, the assertion that John has a new dog is the expression of the appropriate sincerity condition, S's belief that John has a new dog; and if it is the case that he has a new dog, then the conditions of satisfaction of both assertion and belief are satisfied. In light of what precedes, the hierarchical relationship under consideration would be instantiated by that between intentional content and mental state, or intentional content and illocutionary act, as the expression of the mental state.

Criterion 2 – which concerns the aspect under which the lower-level information is construed from the perspective of the higher level – can also be further specified in terms of the logical structure of

3 Why speech acts and the mental states they express have the same conditions of satisfaction would be due to the fact that 'the intention with which the act is performed [...] bestows Intentionality on the physical phenomena', and 'The mind imposes Intentionality on entities that are not intrinsically Intentional by intentionally conferring the conditions of satisfaction of the expressed psychological state upon the external physical entity.' (Searle, 1983, p.27 and pp.169–70)

4 Commissives and directives (Searle, 1983, pp.169–70) also have among their conditions of satisfaction a 'self-referential causality condition'. Thus, to keep a promise, it is not enough to do what one has promised; in addition, one must also do it by way of fulfilling the promise. Similarly, an order is not obeyed 'unless the act the hearer is ordered to perform is carried out by way of obeying the order'.

speech acts. If we take the lower-level element to be propositional meaning P (i.e. the meaning arising from constituent structure, with all referential variables filled in), its primary function[5] as determined by linguistic convention, is to represent a state of affairs. From the higher level, this propositional meaning is seen to be providing the appropriate mental state with an intentional content, thereby also providing the corresponding speech act with a propositional content. In other words, P's lower-level aspect arises from the fact that it interfaces with the external world, while its higher level aspect is determined by its being a component in an intentional structure.

Criterion 3, which is about the directionality of the shift, can be further specified with regard to the nature of the domains of reference involved: thus, the initial domain should be closer to or identical with the external world and the target one further away from it. Corroborative evidence for this directionality – if available – could conceivably be in the form of semantic layering. Divergent thinking, as will be recalled, involves setting aside lower-order information; this entails retaining this information in memory, rather than phasing it out, hence the possibility of there being synchronic traces of this process, in the form of semantic layering. Such layering would consist of two (or more) levels of meaning, with the earlier one(s) – from the standpoint of the process – being presupposed. In addition to being presupposed the earlier layer(s) can be expected to present a higher degree of encoding: the underlying assumption is that the longer a linguistic structure has been used with a certain meaning, the more likely it is to have become conventionally associated with it. The detail of what constitutes indications of varying degrees of encoding will be picked up in due course.

Turning, finally, to Criterion 4, the problem for which the switch to the higher level would provide a better answer is the old problem of ensuring processing efficiency: by providing a shortcut to the interpretation of speech acts – especially one amenable to encoding – divergent thinking allows the interpretative process to cut through a certain amount of inferences it would otherwise have to rely on.

5 Our concern, as may be recalled, is with declarative structures and direct illocutionary acts.

Before proceeding to consider how the case at hand fares by these criteria, let us clarify the role of divergent thinking in relation to the interpretative process and encoding. Among the many ways of performing an illocutionary act, say, an assertion, one of the most economical is probably by means of a linguistic string such as (3), which purports to express a state of affairs:

(3) The flight time was eight hours.

Such a string can pack more information than is predictable on the basis of its constituent structure, in particular, a higher-level meaning to the effect that S believes that the flight time was eight hours; another meaning is that the expression of this belief constitutes a commitment on S's part that the state of affairs is verified. Prior to a systematic use of divergent thinking (with encoding as a possible end-result), calculating the higher-level meaning on the basis of the lower-level string and contextual information (that may not be readily available) would have been extremely inefficient in terms of meeting the demands of the communicative situation. The way divergent thinking helps solve this problem is – as indicated earlier – by providing a shortcut. As an ability to set aside lower-order elements to allow for a sampling of higher-order alternatives, divergent thinking makes it possible, as a matter of routine for:

(a) propositional meaning to be set aside;
(b) the linguistic string conventionally associated with it to become associated with a higher-level meaning, namely a mental state (this would explain that without further contextual information (3) can be taken to express a belief, on S's part, that the state of affairs in P is the case);
(c) the utterance act to become associated with the appropriate higher-level meaning, namely a commitment on S's part to the state of affairs being verified.

As a perceptual strategy, divergent thinking thus allows P to be subordinated to higher-level meanings, that is, to be perceived as the intentional content of both a mental state and an illocutionary act.

Once this perception has become a matter of routine, it would have percolated down to give rise to the perception of an attendant regularity, one involving the use of a certain kind of lower-level string to produce certain types of higher-level meaning. The encoding of higher-level meanings in lower-level strings (or rather the corresponding motor representations, as we shall see) would occur when the appropriate pathways have sufficiently strengthened[6] for the decoding of higher-level meanings from lower-level strings to become automatic.

Direct illocutionary acts with a declarative structure

Assertives and commissives

In virtue of the general requirement about observability, the above criteria are not applicable to the full range of illocutionary acts. Accordingly, we shall have to confine ourselves to direct illocutionary acts expressed by declarative structures. For the purpose of determining whether Criteria 1 and 2 apply, we need a propositional meaning that is immediately accessible and constitutes a self-contained semantic unit, in the sense of being susceptible of a complete interpretation at the lower level. In other words, if our concern is with

6 Where the storage of the representations of facts and their relations is concerned, connection strength among neurons is apparently promoted by a neurotransmitter called 'acetylcholine', issued by the basal forebrain, which 'signals to various structures (including the cortical areas involved in perceptual encoding and the hippocampus) that new representations and/or associations should be stored' (Kosslyn and Koenig 1995, pp.346–7). In the case of the stimulus-response connection subsystem (implemented by the striatum), the reinforcement of relatively direct connections between perceptual inputs and motor programs in motor memory would be largely due to another neurotransmitter, dopamine, produced by a subcortical structure, the substantia negra (cf. Miskin and Appenzeller, 1987; Miskin, Malamut and Bachevalier, 1984; Edelman 1989).

the possibility of a lower-level structure (i.e. a structure conventionally associated with a propositional meaning) becoming encoded with an illocutionary value, we must be able to presuppose the appropriate lower level, one largely determined on the basis of constituent structure, and to have access to it (synchronically). (4) and (5) are typical of what we will be considering:

(4) Prunella is a mezzo-soprano.
(5) I will get you a season ticket.

Both (4) and (5) have a propositional meaning which is immediately accessible and self-contained, and serves to make a direct speech act, in the event an assertive and a commissive. Another way of phrasing the above requirement is in terms of 'propositional content' (as in the distinction between propositional content and illocutionary force. For further details, cf. Austin, 1962 and Searle 1969). As a rule of thumb, the cases that are of interest to us are those whose propositional meaning constitutes an explicit version of the propositional content. In the above two examples, where the illocutionary force is assertion, and promise, respectively, it is clear that the propositional meaning corresponds to an explicit version of the propositional content, /Prunella is a mezzo-soprano/ and /I will get you a season ticket/. Expressives, though also exhibiting a declarative structure, have a propositional meaning which is different from their propositional content. Thus with respect to:

(6) I congratulate you on winning the race.

Searle would have /you won the race/ as the propositional content, with the corresponding propositional meaning being /I congratulate you on winning the race/. Non-declaratives present a similar situation. Consider:

(7) Shut the door.

where the propositional content is /you shut the door/, but the 'propositional meaning' (if 'propositional meaning' is still appropriately used) includes the contribution of an illocutionary marker (the imperative modality). Our assumption in respect of expressives is that, like commands and requests, they will have arisen from a different mode of coding or at least one that is more complex. This point will be taken up further on.

Criterion 1

Consider (4) and (5) again:

(4) Prunella is a mezzo-soprano
(5) I will get you a season ticket.

(4) and (5), it is important to note, are amenable to two levels of interpretation: at the propositional level, they express states of affairs, at the speech act level they function to carry out an assertion and a promise, respectively. Moreover, these two interpretations are linked in the sense that interpreting (4) or (5) as an illocutionary act necessarily entails taking on board the propositional meaning: if the point of an assertion is to commit S to the propositional content being the case, there is no way interpreting (4) as an assertion would preclude taking into account the contribution of the propositional meaning. Similarly, if the point of an act of promising is to place S under an obligation to bring about a certain state of affairs, to interpret (5) as a promise, one would need to specify what that state of affairs is. Inherent in the above is the idea of a hierarchy, with propositional meaning corresponding to the lower level. Further support for this lies in the focus of attention: when propositional meaning is envisaged as a component of a speech act, its relation to the world is not at issue. In the case of (4), the focus is not on the truth value of the propositional content, or its words-to-world direction of fit. Similarly, in (5), the issue is whether S intends to bring about the state of affairs expressed

by the propositional content, rather than the world-to-word direction of fit between the propositional content and the external world.

Thus, inasmuch as it stands in a hierarchical relation to the expression of S's mental state (a belief and an intention), in the case of assertives and commissives, the propositional meaning appears to meet the first criterion. Now, let us see how it fares with the next one.

Criterion 2

Criterion 2 – to jog our memory – is that the aspect under which the lower level content interfaces with the higher-level act should be different from the one under which it was originally envisaged. Consider again the propositional meanings in (4) and (5), /Prunella is a mezzo-soprano/ and /I will get you a season ticket/: at the lower level these propositional meanings are viewed in terms of their relationship to the external world; at the higher level, they are construed in relation to S. In other words, the change in aspect corresponds to a change in representational function: at the lower level they represent states of affairs in the world – albeit future as well as present (or past) – and at the higher level, the intentional content of mental states (belief and intention, respectively), as well as the conditions of satisfaction (in the sense of things required) of the corresponding illocutionary acts.

Having propositional meaning as lower level, however, appears to raise certain problems. There is historical evidence suggesting that propositions expressing mental states would make more likely candidates for further encoding with an illocutionary value. Traugott and Dasher (1987) not only show that the source domain for mental and speech act verbs in English and Japanese is essentially the same, namely mental states (as well as actions, visual perception, spatial relations and vocalization); but, most importantly, that the directionality of the change is from mental to speech act verbs. Thus, with respect to *recognize,* these authors point out (pp.569–70) that:

> although clearly a mental verb in origin (cf. *cognize*) this has been used with the [speech act verb] meaning of *admit* from 11th C. Fr. times on; Souter (1940)

cites a 4th C. Lat. example of *cognoscere* with the same meaning. The same goes for *mitosu* from *mi* – 'see' + *tosu* 'send through' attested as a [mental verb] meaning 'see through to the end, understand', from the 10th C., and a [speech act verb] meaning 'prophesy' from the 17th C.

The hypothesis at hand also appears to be at odds with Searle's view on the nature of speech acts, namely that an illocutionary act arises from the expression of the appropriate mental state on the part of S (cf. Searle, 1979 and 1983). Thus a promise is the expression of S's intention to undertake a future action A, which is beneficial to H, an order is the expression of a desire, on S's part, that H should carry out a future action.

The cases I have been considering differ from those investigated by Traugott and Dasher in one important respect: they are not terms denoting mental states and speech acts sharing the same etymology, but expressions of mental states as part of the meaning of speech acts (e.g. assertions that p include S's belief that p, as part of their logical structure. For more details, see Searle 1983, pp.26–36). This said, given the directionality of the change attested in the verbs at hand – which is from 'propositional' to 'interpersonal' meaning – it would be unreasonable not to consider the possibility that their phrasal counterparts may have followed a similar path of evolution, especially, if Searle's views, which do have a direct bearing on our concerns, also point in that direction.

Are we then to adopt the proposal that the appropriate lower-level element is not a proposition representing a state of affairs in the world but the expression of a mental state, to which that proposition provides an intentional content? In the case of an order, the mental state would be desire, and its intentional object, a future action of H; in the case of a promise, the mental state would be intention, and the intentional object, a future action of S.

This solution, however, also runs into difficulty: if the lower level is the expression of a mental state, how come expressions of mental states have not been encoded with an illocutionary value? Thus it is not inconsistent to say:

(8) I believe John was kissed by Cecilia, but this is not an assertion, or even a suggestion.
(9) I intend to leave you my money, but this is not a promise (or any lesser commitment to that effect).
(10) I want you to be nice to Lee, but this is not an order, or a request (or any attempt to place you under an obligation to that effect).

A possible response hinges on a distinction this objection fails to take on board, one between explicit and implicit expressions of S's mental states. Such a distinction is based on the assumption that only the latter would have been encoded with an illocutionary value. In (8) to (10), what we have got in the first clause is an *ex*plicit expression of S's mental state. *Im*plicit alternatives are given in (11)–(13), again in the first clause:

(11) *John was kissed by Cecilia, but this is not an assertion or even a suggestion.
(12) *I will leave you my money, but this is not a promise.
(13) *You will be nice to Lee, but this is not an order, or a request.

The question this raises is: what underlies the difference between explicit and implicit expressions of mental states that could account for the above difference in encoding? The implicit versus explicit mode of expression corresponds to a difference in illocutionary intention. The intention to express the mental state explicitly constitutes a shift in focus: from the point of the illocutionary act this focus has been redirected to grounds for believing that the state of affairs conveyed by the propositional content corresponds to a state of affairs in the world. Thus, from S's standpoint,

(14) I think he has lost his mind.
(15) I intend to make it up to you.
(16) I want you to do it by tomorrow.

(with the relevant pro-forms in place) would not be appropriately used unless the situation is one in which the reason for believing that the propositional content represents a certain state of affairs needs stressing; such a situation would include a prior act, which, in S's view, has not been successful:

(14a) He has lost his mind.
(15a) I will make it up to you.
(16a) You will do it tomorrow.

In other words, rather than merely convey that S possesses the appropriate mental state, explicit expressions of S's mental states stress that she does, a fact deemed not to have been grasped by H the first time round. For that reason explicit expressions of S's mental states can be discounted as evidence against the proposal that the lower-level meaning either includes a component corresponding to a mental state, or is susceptible of acquiring this additional meaning.

The next point that requires attention is: how come propositional meaning happens to have a 'mental state' as part of its meaning? Thus phrased, this question can be interpreted in (at least) three different ways, only one of which is directly relevant to our purpose. First, it can be about the nature of the mental state involved in the case of each particular speech act (i.e. 'How come we have desire in the case of commands, belief, in the case of assertives, etc.?'). For an answer to this first version of the question, one can turn to what Searle (1983) says about the logical structure of speech acts. More fundamentally, the question could be about why speech acts have the structure they do, in which case Searle again comes in handy, but this time the answer is to be situated at a different level of analysis, and construed in terms of intentionality.[7] Finally there is the 'co-evolutionary' interpretation, which has a direct bearing on our concerns. On this interpretation the question is about why propositional meaning came to imply a subjective point of view (broadly speaking). Thus reformulated, the question is amenable to an answer based on assumptions previously introduced: the switch from propositional to illocutionary level involves the introduction of a subjective angle of

7 Searle's claim that the logical structure of speech acts derives from the structure of intentionality is consistent with a co-evolutionary approach, which would put this link down to a co-opting by language of existing brain structures.

viewing, that of S. As will be recalled, from this perspective, events in the external world, represented in the propositional content at the lower level, become intentional objects of mental states at the illocutionary level.

In connection with propositional meanings and mental states, this time, another question has to be raised, which will be dealt with in due course. It concerns the fact that expressions of the former have been encoded with a content expressing the latter (i.e. 'It is raining' includes /I believe it is raining/, as part of its 'linguistic' meaning). The hypothesis I shall be putting forward is that the expression of the mental state becomes 'internalized' as part of the encoding of the illocutionary value. This internalization concerns certain features of the idealized context of utterance or SN conditions.

But to return to the case at hand, are we then to take the view that the lower level is a propositional meaning that includes a 'mental state' as an implicit content? As pointed out earlier, objections (a) and (b) together make a stronger case for expressions of mental states being the lower level of analysis. This, however, does not entail that such expressions have to be the point of departure of the process whereby illocutionary values become encoded. In viewing propositional meaning as the content of a mental state one is already considering it in terms of its function at the higher level, that is, of its role in the structure of illocutionary acts. Furthermore, in so doing one conflates a (subjective) angle of viewing with its source, a mental state yet to be determined by referring to the relevant illocutionary act.

A more satisfactory account is one that maintains a distinct lower level while seeking to explain what makes it amenable to a certain type of function at the higher level. On this view the lower-level structure is unambiguously a propositional meaning expressing a state of affairs in the world, but one whose intentional object (the state of affairs in question) is susceptible to a certain angle of viewing, the source of which is as yet unspecified. Only at the higher level, where the domain of reference expands to include the speech situation, and a subjective angle of viewing is introduced, does this source become identifiable with S, and the lower level intentional content, with the intentional content of a mental state.

A major difficulty we are facing in trying to identify the lower level, is that our perception of it will be biased by whatever standpoint we adopt. If we view the lower level, retroactively, in terms of its role in the performance of the illocutionary act, it will have a 'mental state' as a part of its meaning. If, on the other hand, we take the propositional level as point of reference, all we are left with is propositional meaning. In other words, either way, one fails to account for how the gap between propositional and illocutionary levels can be bridged.

To get around this difficulty what is needed is a concept of lower level which is both independent of and compatible with the above standpoints. Such a concept begins to emerge if the lower level is construed in terms of different aspects under which it can be envisaged. One involves viewing it in its representational capacity, that is, in relation to the external world; the other requires the standpoint of the illocutionary act, to which it stands in a different relation. These aspects would provide a basis for changes of the kind exemplified at the higher level.

Criterion 3

The first two criteria thus point to propositional meaning (as the implicit expression of a mental state) as the most likely candidate for the lower level content. Does the third one further support this? Is propositional meaning, as specified, closer to the external world? This question can be answered in the affirmative. Consider (4) again. The propositional meaning /Prunella is a mezzo-soprano/ (either on its own, or as the intentional content of a mental state) would appear to be closer to the external world than as part of an illocutionary act. As propositional meaning (pure and simple), it derives its proximity to the external world from the fact that its expression purports to represent a state of affairs in the world. As the intentional content of a belief, it is further away from it, since the corresponding proposition expresses a mental state – a belief – which presupposes the existence of the intentional object, in the event, the state of affairs. In other words, the domain of reference has shifted from the external world to a subjective world. This distance from the initial domain of reference can be expected to increase even further when the propositional meaning becomes the conditions of satisfaction of the propositional content, that is, to the fact that the representation given of the external world conforms to reality (in Searle's terminology, S's commitment is to the words-to-world direction of fit). Though still presupposed (at least, from S's standpoint) the external world constitutes a more distant presupposition, since it is part of a relationship (between words and world) which is itself presupposed and about which there is a commitment on S's part, at the illocutionary level. A similar point can be made, taking H as a point of reference. From the latter's perspective, the external world is also presupposed, though not the direction of fit between words and world. What is presupposed, instead, is the question of whether words and world stand in the relationship presupposed by S (since the purpose of S's commitment is to provide a guarantee for that relationship). From H's point of view, the domain of reference also privileges the interactive level, and presupposes the external world. (5) can be analyzed in a similar fashion: the domain of reference for p (/I will get you a season ticket/) as a propositional meaning is the external world, since it purports to describe an event in that world, albeit a future one. This domain becomes the subjective world of S's mental states, if p is construed as the intentional content of a mental state (in the event, an intention), with the subjective world presupposing the external world. At the higher level, where the domain of reference corresponds to the interactive level, the subjective world, in turn, becomes presupposed (while continuing to presuppose the external world).

Criterion 4

The problem for which divergent thinking provides a more efficient answer is that of identifying the object of communication, in the event, an intentional state, for which the propositional meaning provides an intentional content. How divergent thinking brings this about is by allowing a certain type of attentional shift, from one aspect of propositional meaning to another, that is, from propositional meaning in its representational capacity (in relation to the external world) to propositional meaning as conditions of satisfaction of a mental state and the corresponding illocutionary act. This attentional shift, indeed, results in a shift in domain of reference.

Expressives

Expressives, such as (6), are another class of direct illocutionary acts presenting a declarative structure:

(6) I congratulate you on winning the race.

The reason for dealing with them separately is because their propositional meaning, which includes a description of the illocutionary act, makes a problematic lower level on our terms. While Criteria 1 and 3 could be made to apply, the failure of Criterion 2 – by far the most important – to do so, alerts us to the possibility that a different type of coding process may be involved: /I congratulate you on winning the race/ cannot work as propositional content of the act of congratulation. To complicate matters further, it is unclear what the correct propositional content should be of /I rejoice in your winning the race/ or /you won the race/, as Searle would have it.

This prompts the following two questions:

(a) Are we dealing with a totally different mode of coding, as in the case of directives, where there is an illocutionary marker, or, is divergent thinking still involved?
(b) If the latter, how does it fit in with the appropriate coding process?

Given our working assumptions, arguments in support of the involvement of divergent thinking in the coding process crucially depend on the availability of a likely candidate for the lower level. Such a candidate must be is accessible synchronically, and stands in a certain relationship to the higher level. The appropriate lower-level content, as will be recalled, must (i) remain, with a presuppositional status; (ii) provide the higher level with a plausible propositional content (through the mediation of an attentional shift that highlights its link to the higher level); (iii) possess a domain of reference that is closer to the external world than its higher-level counterpart; the domain corresponding to the latter must also be more abstract in a way that is consistent with divergent thinking.

If no plausible candidate for the lower level is accessible synchronically, as seems to be the case with expressives, the identification of an earlier candidate may prove difficult in view of likely semantic changes undergone by relevant linguistic elements.

Before we return to this issue, let us consider Searle's definition of expressives (1979, p.15). Searle's view is that 'The illocutionary point of this class is to express the psychological state specified in the sincerity condition about a state of affairs specified in the propositional content'. Thus, in (6), the purpose of the act is to express S's pleasure on the occasion specified in the propositional content, that is, of H's winning the race. Furthermore, expressives are distinctive in not having a direction of fit, since the purpose of the act is to express S's psychological state, and the propositional content purports to represent an external state of affairs. Another way of putting it is that, unlike what happens in other cases, the propositional content is presupposed and does not specify the conditions of satisfaction of the act.

If – as I propose to do now – one were to construe expressives in a different way, certain parallels with other illocutionary acts, in particular with assertives, begin to emerge, which have a direct bearing on our concern with the lower level. Instead of assuming that propositional contents represent objective states of affairs in the world only, suppose one were to extend the notion to subsume subjective states of affairs, or psychological states, as well, (after all, from an ontological standpoint,[8] the latter, arguably, are also part of the world). What becomes immediately obvious is that like assertives, expressives can be seen:

(a) to have a direction of fit, from words-to-world, (where 'world' includes minds);

8 Cf. Searle 1992, p.13, for a distinction between the epistemological and the ontological sense of the subjective.

(b) involve S's commitment to the words-to-world direction of fit. In this case, as in the others, the propositional content determines the conditions of satisfaction of the act.

Another crucial aspect of expressives is that they possess a manipulative dimension of a sort, which is difficult to ignore. Unlike directives, the purpose of expressives is not to get H to bring about the direction of fit between world and words. As far as the responsibility of bringing about the direction of fit about is concerned, they are rather like assertives, in that this is achieved via S's commitment to the propositional content being the case. Expressives, however, are manipulative in a sense that assertives are not, as their illocutionary point might be described as one of promoting social cohesion, of mending the rapport between interlocutors, which has been, or is likely to be damaged, should S fail to respond in an appropriate manner to the event presupposed in his utterance. Thus, S's failure to congratulate H on winning the race may cause resentment on the latter's part. Similarly, if I were to sit on your hat, failure on my part to express regret, as in:

(17) I apologize for sitting on your hat.

may be regarded as a breach of civility. If the purpose of assertives is to commit the speaker (to varying degrees) to something being the case (Searle 1979, p.12), that of expressives, in my view, goes one step further: in committing himself to having the appropriate psychological state, S is performing a social act of the kind described above.

One could certainly take the view that the point of expressives is to express the psychological state (see Searle's definition), and that the social act is being performed indirectly. I see at least two difficulties to that. First, Searle's definition would also apply to a situation where H has no part in the winning, as in:

(18) I am overjoyed I won the race.

The second difficulty is that the act of congratulation is not carried out in (6) the way it is in:

(19) I am so glad you won the race.

which is a clear case of indirect speech act. To compound matters further, *to congratulate, to apologize, to regret,* which Searle takes to be paradigms of expressive verbs, denote social acts carried out via S's commitment to the relevant psychological states being the case, not just the expression of those states.

In short, where Searle views the purpose of expressives as one of communicating S's psychological states on a certain occasion, I see that purpose as one of committing S to the truth of the propositional content (in the extended sense) by way of performing a social act. This commitment, which is parallel to S's commitment to the truth value of p in the case of assertives, is what would, under certain conditions, constitute an act of congratulation, apology, or regret, as the case may be. When I apologize to H for sitting on his hat, I do not merely express regret for doing so, I also do two further things:

(a) I commit myself to the truth of the propositional content (as an assurance of my sincerity), or rather, purport to do so, since this is not the ultimate point of the act;
(b) I do (a) by way of reparation to H's feelings.

Point (b) is where expressives diverge from assertives to present a clear resemblance to declarations, of which extra-linguistic examples are declaring war, adjourning a meeting, and linguistic ones, stating, ordering, promising by way of declarations (cf. 'I state that all vaccines based on bovines material have been destroyed.', 'I order you to tell the truth', 'I promise to do it tomorrow'). A declaration, by definition, creates new facts, and a linguistic declaration is a speech act whose point is to create a new fact corresponding to the propositional content. Alternatively, '[an] utterance is a declaration if the successful performance of the speech act is sufficient to bring about the fit between words and world, to make the propositional content true.' (Searle 1989, p.549) The sort of apparatus Searle sets up for performatives in general would appear to work for expressives. Thus, when I say:

(6) I congratulate you on winning the race.

my utterance creates a new fact corresponding to the propositional content (/I congratulate you on winning the race/).

We are now in a better position to return to our initial question, which was: what would an earlier candidate for the lower level look like? If the parallel with assertives is something to go by, just as assertives are performed via utterances expressing states of affairs in the objective world, so earlier expressives would have been performed via utterances expressing states of affairs in the subjective world, or psychological states. This scenario appears to be consistent with the derivational history of a number of expressive verbs or expressions. Thus, the *Oxford English Dictionary* (O.E.D.) entry for *congratulate* (from Latin *congratulari*, to rejoice along with another person) lists 'to rejoice along with another person', 'to express joy, pleasure or satisfaction on an event or circumstance', as prior to the illocutionary meanings of 'to address a person with expression of joy or satisfaction on an occasion considered fortunate', 'to offer congratulations'. *Féliciter*, as in:

(20) Je vous félicite d'avoir remporté une victoire écrasante.
I congratulate you on winning this crushing victory.

exhibits a similar pattern. The *Trésor de la Langue Française* (T.L.F.), lists 'rendre heureux', 'féliciter avec vue d'assurer une personne de la part qu'on prend à la joie de quelque chose', as earlier than 'faire des compliments, congratuler'. *Thank*, according to the O.E.D. meant 'thought, favourable thought', 'kindly thought or feeling entertained towards any one for favour or services received, before 'grateful acknowledgement of a benefit or favour'. Not all expressives, however, exhibit this pattern. The O.E.D. does not list for *apologize* any early psychological state meaning on its own. Such a meaning only occurs as part of the illocutionary meaning, 'to acknowledge and express regret for a fault without defense, by way of reparation to the feelings of the person affected'. Note, however, that once *apologize* has become an expressive verb, it falls into step with other expressives, in having a similar logical structure.

Given the nature of diachrony, it is hardly surprising that expressives do not all have the same derivational history. The real difficulty lies elsewhere: what is presented as a chronological list of attested meanings for a given linguistic expression would perhaps be better described as a list of attested occurrences, whose meanings had to be worked out on the basis of the linguistic context (with the aid of earlier lexicons, where available). Traugott and Dasher express similar reservations in the study mentioned earlier:

> It is often difficult to tell whether we really have a speech act verb [...]. When the first clear instances of speech act verb meanings for earlier mental verbs do appear, they often occur in passive sentences (i.e. not in sentences of the prototypical performative type *I hereby verb that*. (Traugott and Dasher 1987, pp.570–1)

A further difficulty is in distinguishing between linguistic meaning and meaning induced by the situational context.

Difficultie*s* aside, the scenario proposed earlier for prototypical expressives is still incomplete. What has not been taken on board is the fact that those expressives are associated with 'performative' utterances. Thus, even if at an earlier stage an act of congratulation, (cf. (6)), was conventionally performed via an utterance with a propositional meaning expressing S's psychological state (i.e. something to the effect that he rejoices in the fact that H won the race), this does not tell us why the lower-level structure includes a description of the speech act being performed in saying it.

In what follows, my first concern will be how the gap just mentioned can be filled by appealing to Ducrot's views on performativity. Then I shall return to my original question, which was whether divergent thinking still has a role to play in the emergence of the expressives at hand.

But first we need a summary of what has been said so far:

(a) At an earlier stage, to congratulate someone upon a happy occasion O, one would have used a lower-level string (LL), whose propositional meaning expresses S's pleasure on O, or rather S's sharing H's pleasure on O. (LL would have meant something along the line of: 'I rejoice with you at O.')
(b) Through persistent use of LL to perform acts of congratulation, the utterance of LL came to be conventionally associated with the performance of an act of congratulation.
(c) [is where we have a gap]
(d) Corresponds to modern usage, where an act of congratulation can be carried out by means of a verb denoting an act of congratulation, as in (6).

Ducrot's account of stage (c) (1975, pp.85–6; and 1984, pp.117–48) involves a type of derivation called 'délocutivité élargie' ('extended delocutivity'). As a quick reminder, delocutivity, as construed by Benveniste (1966, pp.285–7), is a lexical process of derivation originally postulated to explain the emergence of verbs such as *bisser* (to encore, in the sense of 'to ask for an encore') from 'Bis!', used conversationally, at first, then conventionally, to ask for an encore. *Salutare* (to greet) and 'Salus!' would stand in a similar relationship. Unlike a morphological derivation linking one word with another, delocutivity is concerned with how meaning evolves, and, more specifically, with how a lexical value can arise from a pragmatic one. Ducrot's view on performative verbs is an extended version of this hypothesis. Consider *promettre* (to promise), which may be regarded as parallel to *congratulate*, since both can denote an illocutionary act and occur as part of a string whose utterance is conventionally used to perform the act they refer to (e.g. 'I promise to come.' and 'I congratulate you on winning the race.') According to Ducrot, the contemporary meaning M2 of *promettre* – as that of similar verbs (e.g. *ordonner* (to order), *permettre* (to allow)) – results from a process such as the following: *promettre* at an earlier stage had a meaning M1 (*mettre en avant* (to put forward)), which did not denote an illocutionary act, and was used, with that meaning, in the first person of the indicative present (or 'dramatic present'), to perform an act of promising. In time, 'Je te promets (M1) de faire A'

('I promise to do A') came to be conventionally associated with this usage. This, in turn, gave rise (via delocutivity) to a new meaning M2, for the verb, one denoting the illocutionary act that can be performed in uttering 'Je te promets de faire A'.[9] The various steps involved in a delocutive derivation, which provide a possible scenario for the gap corresponding to stage (c), can be summarized as follows:

Step 1: We have a verb V with a meaning M1, which makes it possible for it to be used in the first person of the 'dramatic present' to perform an indirect illocutionary act.

Step 2: The use of sentences of the form 'I V', where V has the meaning M1, to perform that illocutionary act becomes conventional.

Step 3: Through its occurrence in the above type of sentence, V acquires, via delocutivity, a meaning M2 in virtue of which it now also denotes the illocutionary act at hand.

So, if we were to assume that *congratulate* resulted from a delocutive process, this is what one would have for Stage (c): once LL ('I rejoice with you at O') has become conventionally associated with the performance of an act of congratulation, *congratulate*, which hitherto had the meaning M1 ('to rejoice with someone on a certain occasion'), acquires a meaning M2, whereby it now (also) means 'to perform the act being performed in uttering LL'. It is this latter meaning that *congratulate* has in (6).

We now return to the role of divergent thinking in the coding process of expressives, especially in relation to delocutivity. In light of what precedes, a more comprehensive scenario can be envisaged for expressives like (5), for which there appears to be an attested earlier lower level. But first let us repeat an earlier version of it:

9 *Promettre*, as Recanati (1981, pp.124–5) points out, may not be the best choice of examples imaginable, since Latin *promittere* already meant *to promise* (i.e., M2). This point taken, given the nature of diachrony, no derivation can reasonably be expected to present an absolute level of generality. As long as Ducrot's proposal works for a range of cases, and is compatible with the working hypotheses at hand, it is a possibility worth considering.

(a) At an earlier stage, to congratulate someone upon a happy occasion O (e.g. on winning a race), one would have used a lower-level string LL (in the first person singular of the dramatic present), whose propositional meaning expresses S's pleasure on O, or rather S's sharing H's pleasure on O. LL at this stage already included a verb *congratulate* with a meaning M1 (M1 = 'to rejoice along with someone on something').
(b) Through persistent use of LL to perform acts of congratulation, the utterance of LL came to be associated with the performance of an act of congratulation.
(c) Once the use of LL to carry out acts of congratulation has become conventionalized, *to congratulate* acquires via delocutivity a meaning M2, 'to perform the act being conventionally performed in uttering LL'.
(d) LL is reinterpreted as including *congratulate* with meaning M2. Hence the emergence of a performative interpretation for 'I congratulate you on winning the race'.

If there is an attested lower-level structure of the appropriate kind, divergent thinking can plausibly be brought in to play the same role as in other cases. Thus, stage (a) and (b) could be further specified in respect of the process involved in going from propositional to illocutionary meaning and what forms the basis for the subsequent recoding process.

In Stage (a), divergent thinking would be the underlying capacity that makes possible the shift from propositional to illocutionary meaning: in order to go from one to the other, the focus of attention must be redirected from one aspect of the propositional meaning to another, that is, from propositional meaning in a representational capacity with respect to the world, to propositional meaning as standing in a hierarchical relation to S's illocutionary intention. In other words, the shift is from the expression of S's pleasure in seeing H win (as a representation of a state of affairs in the 'world'), to the intention to perform an act of congratulation. As a result of this shift, which involves viewing the propositional meaning from a subjective angle, this meaning 'becomes', at the higher level, the propositional content of the speech act, whose conditions of satisfaction it expresses. This for Stage (a). Stage (b) – which is when the utterance of LL becomes conventionally associated with the performance of an act of congratulation – can be further described as follows: through persistent use of LL (or rather structures such as LL) to perform acts of congratulation, the 'active link' between LL and S, which all similar structures share, becomes encoded with the illocutionary value. (This way of construing the coding of illocutionary value is more accurate than simply talking about the utterance acquiring an illocutionary value.) Regarding stage (c), if we go along with the delocutive hypothesis, this is what could have happened: once *congratulate*, following Stage (b), has acquired a meaning M2 (= to perform an act of congratulation), both lower and higher levels are reinterpreted accordingly (= Stage (d)), hence /I congratulate you on winning the race/ as a lower-level content, is now taken to denote the speech act its utterance serves to perform, while as a higher-level content, it represents the conditions of satisfaction of the act, which are that S is hereby performing this act. Stage (c) is thus Stage (a) revisited, with the same type of switch operating, albeit on a different pair of contents, since the lower level has changed. Without divergent thinking one would have been at pains to explain how S's commitment to a certain mental state leads to the creation of a new fact, namely, an act of congratulation. Divergent thinking makes it possible to think of the original lower-level structure as already conventionally associated with the appropriate illocutionary value, which convention is carried over to the 'post-delocutive' lower-level structure.

What advantages are there in of viewing speech acts in terms of divergent thinking and attendant parameters? The existence of a break in continuity between propositional and illocutionary meaning has long been recognized, but, to my knowledge, never looked into. In any case, from our standpoint, this break is construable in terms of a switch to a higher level of abstraction, one made possible by divergent thinking. What divergent thinking helps clarify is how the switch is carried out in terms of attentional shift, aspectual changes, levels of abstraction, changes in viewpoint, and changes in domains of reference. Being able to provide this level of specification also enables us to identify the active link between the utterance and S as a more plausible basis for the coding of illocutionary values than the utterance as the end product of the act of utterance. But there is yet a further advantage: inasmuch as divergent thinking, as an underlying capacity (as opposed to a form of derivation), is independent of the diachronic perspective, the conditions required for deciding its presence (which are construable in terms of the relationships between lower- and higher-level contents and changes in domains of reference) provide additional criteria, which can be used concurrently with or independently of historical data. A final advantage is the possibility this opens up in terms of grounding.

An issue the above may prompt is whether the adoption of delocutivity constitutes a wholesale commitment to Ducrot's stance vis-à-vis performativity. Ducrot takes the view that:

(a) The possibility of performing an order in uttering a structure like 'I order you to do it!' is a matter of convention.
(b) It is an illusion to think that one can perform an illocutionary act by describing or asserting that one is doing it. This illusion of performativity ('l'illusion performative') Ducrot imputes to a common practice in Philosophy of Language, that of using every day language as a metalanguage.

I have no problem with the view expressed in (a). Acknowledging that one is dealing with a convention does not preclude asking what other factors may have contributed to the persistence of such prior behaviour leading up to encoding. One such factor has to do with the possibility of a connection between meaning and other forms of intentionality (e.g. perception, intention, mental states) – a connection Searle (1973) postulates, which also fits in with a co-evolutionary approach (since the latter assumes that language, as a latecomer, would have co-opted existing structures and strategies best suited to its purpose). Another factor concerns the underlying capacity that makes it possible for interpretative processes to go from propositional to illocutionary meaning, in repeatable and consistent ways across individuals over extended periods of time.

In light of what precedes, some circumspection is required in handling point (b). If the performative thesis is construed in terms of means-to-an-end from the standpoint of a user reflecting on the means available to perform a speech act, then I could not agree more with Ducrot: to say that she does so by describing the act in question does indeed fall short of providing a proper answer and is better nipped in the bud by invoking a convention. However, if the question is really about what underlies the emergence of that convention, then I think it deserves further consideration. Although what I have in mind is totally different, it is worth taking a quick look at Searle's handling of performatives (1989, pp.555–6). The way Searle states the problem clearly shows that his primary concern is with underlying factors. Thus, the problem of:

> how the literal utterance of certain ordinary indicative sentences can constitute, and not merely describe, the act named by the main verb (or some other performative expression) in that very sentence [...] turns out under investigation [...] [to be] the same question as how the literal utterance of those sentences can necessarily manifest the intention to perform those acts.

Why this should be so lies in the fact that 'for such acts, the manifestation of the intention is constitutive of the performance.' The solution to the problem then is: just as 'any illocutionary act can be performed by an utterance, because the utterance can be the manifestation (and not just a commitment to the existence of the relevant intention)', so can they be performed by 'a performative utterance, because the performative utterance is self-referential to a verb which contains the notion of the intention which is being manifested'.

What emerges from this discussion, however, is that from the linguistic standpoint neither Searle, nor Ducrot, believes there is more to performativity than conventions and institutions (although each reaches this conclusion via a different route). Furthermore, Searle – unless I am mistaken – does not make any claim to the effect that the utterance of a performative sentence is sufficient successfully to perform the act being referred to. For this to occur, a number of contextual conditions need to obtain, in particular that S has the intention to perform the act in question, not just to manifest that intention. From the philosophical standpoint, on the other hand, the original question can be rephrased in terms that no longer make it vulnerable to Ducrot's objection.

By way of closing this section, I propose to pick up a couple of loose threads. With regard to the question of self-referentiality, from our standpoint there are two levels of self-referentiality. We said earlier that the basis of the coding of the illocutionary value is the process whereby S uses the lower-level structure or rather the active link between S and LL (this, incidentally, seems consistent with Searle's notion that the utterance act can be the manifestation of S's intention). If that is the case, once the coding has taken place, each utterance of LL will constitute a reference – albeit an indirect one – to the process that links LL to S. The other level of self-referentiality corresponds to the explicit reference we get with a performative verb. In addition to this, another difference can be made between these two types of reference: it concerns the aspect involved. In the former case the relevant aspect is the active link between S and LL, in the latter, the reference is to the act taken as a whole. In terms of domains of reference, the former pertains to a subjective domain, the latter to the context of utterance as a subdomain of the external world. From this standpoint, all illocutionary acts are self-referential in the first sense, and only those performed via performative sentences will also be self-referential in the second sense.

Chapter 2

SN conditions and the internalization of environmental constraints

Conditions for the success and non-defective performance of speech acts

This chapter is centred on our second case of language adaptation to processing needs. As is well known, 'lower-level'[1] structures whose utterance is conventionally associated with an illocutionary act appear to have been encoded not only with an illocutionary value, but also with a number of 'conditions for the success and non-defective conditions of speech acts' (SN conditions). Thus, denial of one such condition results, in (1) to (8), in the cancellation of the speech act preceding 'but':

(1) *You will shut the door on leaving, but I know you can't do it.
(2) *You will shut the door on leaving, but I know I have no right to expect you to.
(3) *I promise to do it, but I don't think it is feasible.
(4) *I promise to do it, but, as you know, it is hardly in your interest.
(5) *I advise you to do it, but it is not in your interest to follow my advice.
(6) *Bostridge is a counter tenor, but it is sheer speculation on my part.
(7) *I have to catch a plane on Sunday, but you know that already.
(8) *I will make sure it gets in the post by tomorrow, but I have no intention to do so.

1 'Lower level structures' (or LL) is shorthand for 'structures conventionally associated with a lower-level or propositional meaning, by linguistic convention'.

These examples prompt the following questions:

(a) Why, in addition to the encoding of illocutionary value should there be a need for further encoding, involving, this time, SN conditions, or an idealized context for the performance of the speech act?
(b) What sort of process are we dealing with, and how does it fit in with the apparatus introduced so far?

As suggested earlier, idealized contexts would have become encoded in response to an increasing need for processing efficiency, or rather,[2] once under way, the process would have been reinforced and completed in view of its contribution to processing efficiency. Processing efficiency at the text level is to be construed in terms of goals specific to the communication situation, in the event that of ensuring, insofar as possible, that H will respond in the expected way; thus, H is expected not only to identify the encoded information, but also to accept the obligation to fulfil the conditions of satisfaction of the act, in manipulative cases. In-built idealized contexts would go some way towards helping achieve these goals. Indeed, their effect would be one of inducing in H the formation of a mental context in which the speech act is both successful and non-defective. Inasmuch as that context has become part of the meaning of the utterance being processed, this ensures the latter's automatic decoding, as a certain type of illocutionary act. However, although idealized contexts cater for the proper identification of the speech act, they do not invariably secure the fulfilment of its conditions of satisfaction in manipulative cases: the successful identification of an utterance as an order does not necessarily entail H's acceptance of the obligation to carry it out, despite the fact that the in-built context presents the act as non-defective in every respect. In some cases, there may even be a degree of constraint on response: when caught unaware by a request – especially of a seemingly innocuous kind, such as, 'What time is it,

2 This rephrasing is crucial if one wants to avoid introducing any intentionality in changes that, arguably, are not intentional.

please?’ – one tends to respond appropriately (in the sense of providing the expected response). This, however, constitutes no reliable indication of compliance with the obligation at hand, as responding inappropriately would require an effort of attention or some priming to that effect arising from previous experience.

Encoded idealized contexts and the internalization of environmental constraints

Support for the idea that idealized contexts became objects of encoding in response to an increasing need for efficiency is to be found in the parallel it presents with genetic assimilation (Deacon 1997, p.324). Apparently, in cases where survival depends on how quickly and appropriately organisms can respond to external conditions, natural selection will tend to replace flexible adaptive responses to continuing environmental constraints with genetic predispositions. This is reflected in a progressive canalization of adaptive responses, which become less dependent on external signals and more biased by genetic inheritance. Consider the example of the evolution of the alarm calls in vervet monkeys cited by Deacon (1997, pp.330–1). What appears to have enabled those calls to become genetically assimilated is the correlation between a distinctive invariant set of features in the alarm-calling context and ‘consistent differences in neural architecture’. For each type of alarm call, four types of stimulus-response parameters must remain invariant. ‘The first invariant feature is the motor programs and sensory templates of the vocalizations themselves [...] Second are the invariant sensory characteristics of the stimuli’, that is, the general attributes that distinguish one vervet predator from another, which are consistently instantiated by distinct neural circuits. Such attributes include:

> overall silhouette, size, movement patterns [...] The third invariant feature is the emotional arousal state elicited by both calls and real predators. The startle and fright responses elicited by the sight of a predator are, of course ancient and consistently localized features in the nervous systems of most species of

> vertebrates [...] The fourth, and final factor, is the invariant nature of any successful escape response.

In the case of speech acts, the external conditions most likely to initiate an appropriate response in H would be found in a context of utterance where SN conditions obtain. The 'internalization' of these conditions as distinguishing features of specific types of speech act would be a first step towards creating an automatic connection between speech acts and responses that will fulfil their conditions of satisfaction, as the in-built context would ensure a proper identification of the speech act.

Distinguishing between SN conditions

Before we take a closer look at this hypothesis, it is necessary to ask whether all SN conditions are similar with regard to their susceptibility to encoding, since the sincerity condition stands apart: from H's standpoint, all these conditions (excluding the essential condition, which amounts to saying that there is a convention) could qualify as external; however, of all the sequencings involving the denial of SN conditions, the one involving the sincerity condition (cf. (8)) is by far the least acceptable. Furthermore, the utterance of the appropriate lower-level structure in the performance of a speech act is the manifestation of the sincerity condition. In other words, it would appear that our hypothesis about the co-option of a strategy similar to internalization does not apply to the sincerity condition.

But what kind of scenario does this hypothesis entail? To say that the (potential) utterance of a given lower-level structure LL becomes conventionally associated with an illocutionary act means that the pathways subserving the use of LL to carry out the act in question become routinized, which entails – among other things – a direct linkage between the corresponding motor representation and the illocutionary value. Now the question is, what form does that illocutionary value take? If – as Searle proposes – the reason that the utterance of certain types of sentences constitutes the performance of the corresponding speech act lies in its being the manifestation of the corresponding mental state, then the answer to the question is that the encoding of the illocutionary value is no other than the encoding of the mental state value. Thus, the process of using a proposition expressing a future act of S under certain conditions constitutes the expression of S's intention.

Having established that the sincerity condition could not be regarded as an external condition that would have undergone internalization, we are left with preparatory conditions about the relevance of the speech act, its legitimacy, the possibility that its conditions of satisfaction will be met. Another look at (1) to (7) will tell us that (4), (5) and (7) are about relevance, (2) and (6) about validation, (1) and (3) about physical possibility.

Two objections

Before attempting to make a case for the possibility of a strategy similar to internalization, we must consider two possible objections. One is: Why can't we be content with a linguistic approach? The other concerns the need to bring in internalization to explain the encoding of SN conditions, when divergent thinking could conceivably do the job.

A linguistic approach to in-built contexts would run along these lines: at an earlier stage, whenever S made a speech act, by means of LL, he would imply that the context is one in which all SN conditions obtain. And this implicature would normally be successfully retrieved by H, because both S and H would share the knowledge of what such a context should be like and each knows – albeit not necessarily consciously – that the other knows that he knows. When the use of LL to perform the speech act in question became conventionalized, so would – albeit to varying degrees – the various inferences about the context. An approach such as just outlined is not fine-grained enough to serve our purpose, since it simply makes reference to encoding processes without offering any further specification. But quite apart from a lesser concern for grounding, a linguistic approach is not in a position to account for aspects of meaning it does nevertheless recognize, namely, in-built contexts, or the fact that speech acts are a form of action. Even if one were to leave aside the specification of underlying processes, invoking conventionalization both in the case of illocutionary value and that of preparatory conditions runs into an obvious difficulty: while it is plausible to assume that persistent use of the same types of LL to perform the same types of speech act will result in the appropriate illocutionary value becoming conventionally associated with these LLs, contexts do not present the same degree of invariance. For one, it is difficult to imagine that, unlike us, earlier speakers will only perform a speech act if all the SN conditions obtain in the actual context (surely, then just as now, not everybody who gives a linguistic order has the right to issue a bona fide one). Furthermore there is the problem of how H can go about verifying that all conditions actually obtain, especially those that are about S's mental states. To get round this difficulty, one would have to suppose that, whatever the actual context, H's perception of it would tend to be distorted or corrected by the use of a prototypical situation. But to envisage this eventuality is already to go beyond conventionalization. To this should be added the fact that on a linguistic approach, all SN conditions, including the essential condition, would be handled in the same way, which is far from satisfactory in view of the special relationship in which the latter stands to the illocutionary value.

The second objection was about the need to bring in internalization, when divergent thinking – one would have thought – should have amply sufficed: after all as a capacity to seek allocentric associations divergent thinking does not merely shift attention to higher-level alternatives, but also to alternatives on the same level.[3] On this scenario, once divergent thinking has provided an interpretative shortcut between the LL associated with a propositional meaning and the illocutionary meaning, it would again intervene, but this time to widen the scope of attention; in so doing it enables the latter to encompass peripheral entities that are on the same level, namely elements of the context of utterance. In this case, as in the previous one, divergent thinking would thus provide an interpretative shortcut, thereby making it possible for a direct link to become

3 Cf. Deacon 1997, p.265: 'The prefrontal cortex helps us inhibit the tendency to act on simple correlative stimulus relationships and guide our sampling of alternative order sequential or hierarchical associations.'

established between the utterance of LL and contextual values. As it stands, this objection appears to reflect a concern for simplicity and elegance which is at odds with our type of approach, where a convergence of factors – some of which purely fortuitous – is the norm. An account in terms of divergent thinking and one involving internalization need not be mutually exclusive, as divergent thinking and internalization can be regarded as complementary strategies contributing in different ways to processing efficiency. The contribution of divergent thinking – as we already know – lies in privileging a mode of thinking that facilitates allocentric associations, that is, the linkage between linguistic structures and higher-level meanings to which they are not associated in virtue of a linguistic convention. In this sense its role is to provide an interpretative shortcut between a given structure and a meaning it is used to express, where the retrieval of the latter from the former would normally involve a certain amount of inferencing. By contrast, internalization is a complex process whereby the availability of external conditions becomes less dependent on the environment and more dependent on brain processes, thereby enabling hitherto flexible behaviour to become automatic. Having organisms depend on the occurrence of a whole range of external conditions before they respond appropriately would be costly in terms of survival. As a strategy – internalization or a tendency to that effect – would greatly facilitate the encoding of external conditions in a way that is different from divergent thinking. Where divergent thinking, in shifting attention from egocentric to allocentric association, makes it possible for a shortcut to be hit upon, internalization, as a strategy, creates a precedent for contextual features, rather than other kinds of allocentric associations, to be encompassed in the widening scope of attention brought about by divergent thinking.

Speech acts and 'genetic assimilation'

We now return to our original question, about the well-foundedness of viewing the encoding of SN conditions in terms of a process similar to the internalization of external conditions. Inasmuch as these conditions derive their raison d'être from speech acts, this question cannot be dealt with separately from the issue of the eligibility of speech acts to genetic assimilation, where 'genetic assimilation' is to be taken in the sense of 'an evolutionary trend toward increasing developmental constraint', rather than 'the one-to-one replacement of learned adaptations with instinctual counterparts' (Deacon 1997, p.324).

As part of his discussion of the susceptibility of language universals to genetic assimilation, Deacon (1997, pp.329–39) sets out two conditions a language feature must fulfil in order to have an impact on brain evolution. One is that 'it must remain invariable across even the most drastic language change possible'; the other is that 'this persistent aspect of language must also impose consistent invariant demands on neural processes', in the sense that the neural computations it requires must consistently be performed in some local structure in human brains, or alternatively exhibit invariance in the way they are distributed within different brain structures. This second criterion, which severely restricts what can be eligible to participate in Balwinian[4] evolution, most crucially exclude from this category logical operations pertaining to the 'deep structure' of grammar, which have often been proposed as universals of language. Indeed these operations vary considerably in their surface implementations from one language to another, and invariance in surface structure is, among other things, what determines processing invariance. Why these operations should be so cut off from their surface implementation, Deacon attributes to the non-correlative nature of symbolic reference: insofar as languages 'exhibit minimal correlation between words and their referents', it is to be expected that 'grammatical operations have even less of a correspondence to things in the world'.

4 'Baldwinian evolution' refers to a slight modification of Darwin's theory of selection proposed by the American psychologist James Mark Baldwin. 'Baldwin suggested that learning and behavioral flexibility can play a role in amplifying and biasing natural selection because these abilities enable individuals to modify the context of natural selection that affects their future kin.' (Deacon 1997, p.322)

The one language feature, which, in Deacon's view, eminently satisfies both criteria for genetic assimilation is the learning bias that underlies symbol acquisition, and makes it possible for humans to 'overcome the mnemonic and attentional threshold that tends to prevent sets of rote indexical associations from becoming recoded into symbolic systems of associations'. Symbols acquisition, which is arguably required by all languagelike behaviour, 'largely depends on processes that are carried out in the prefrontal cortex'. Deacon singles out symbol use as the 'prime mover for the prefrontalization of the brain in hominid evolution.' Language has thus 'given rise to a brain which is strongly biased to employ the one mode of associative learning that is most critical to it'.

Two other important sources of genetic assimilation appear to be the demands made by the use of language on neural systems involved in the production and perception of speech, on the one hand, and the processing of information, on the other. Given the amount of phonemes and information being produced during linguistic communication, processing efficiency requires that a significant part of these operations should be automatic. (Deacon 1997, pp.336–7)

A yet further source is provided by learning predispositions,[5] be it in the sense of 'a predisposition to learn some particular class of associations', or a predisposition 'to attend more closely to certain stimulus features during learning, or merely a bias in motivation toward certain realms of learning'. Thus, besides the genetic assimilation of neural templates (that arose from sensory and behavioural invariance towards relevant features of the adaptive context), one should also envisage the possibility that certain invariant features of the learning context itself have become internalized.

With the above criteria in mind, we can now return to speech acts and consider their susceptibility to genetic assimilation. Assuming that speech acts are a universal feature of language, and that all language-like behaviours, or at least those that are of a manipulative kind,

5 In Deacon's view (1997, pp.337– 8), since learning is based on links and biases among separate neural systems, learning biases would be represented in brains in a different way from, say, modular abilities such as those involved in speech sound analysis.

possess them, what is invariant about them cannot be their surface structure, which is different from one language to another, and subject to change within the same language. The first possible source of invariance lies in the use of divergent thinking. As will be recalled, divergent thinking would have been responsible for what I have called a 'further symbolic recoding', which marks the passage from propositional to illocutionary meaning: going from one to the other requires an attentional shift, which sets aside lower-order associations to seek higher-order ones. In other words, as a case of intensive further application of this capacity, the use of speech acts could have an important role to play in reinforcing its impact on brain evolution.

As it stands, this proposal is open to an objection, which is difficult to ignore: over time the persistent use of certain syntactic structures in performing speech acts, would have become conventionalized; in other words, where at an earlier stage one would have had an inferential process involving divergent thinking – as an attentional shift was required to connect propositional to illocutionary meaning – at a later stage, this inferential process would have been replaced by an automatic link, which no longer required it.

In addressing this objection we must be careful to distinguish between three different uses of divergent thinking. Divergent thinking, as will be recalled, is the capacity to inhibit the tendency to follow up correlations arising from egocentric information, and to shift attention to alternative associations. So far, the only use we have been concerned with is the one involving the shift of attention to higher-level alternatives and exemplified in the shift from propositional to illocutionary meaning. However, this allocentric use presupposes the use of egocentric information as a frame of reference, which implies that this information – instead of being ruled out – merely recedes to the background, where it can be retained: this is consistent with the fact that interpreting an utterance encoded with a speech act does not entail cancelling its propositional meaning, which, moreover, provides the propositional content of the speech act. Inherent in the above use of divergent thinking (let us call it 'symbolic'), there would be a more basic, 'divergent function', that can still operate when an automatic process takes over from the symbolic function. A further use is one involved in the selection of the illocutionary force, or rather, in generating contextually plausible alternative interpretations, as part of that selection process. This 'interpretative' use of divergent thinking[6] [7] can be seen at work in the derivation of indirect speech acts (cf. the use of 'He went off like a shot', as a warning of impending disaster); but also in the selection of individual speech acts within the same class (e.g. 'My dog makes speech acts all the time', which may be interpreted as a statement, a boast or a complaint depending on the context). In support of these distinctions one can refer back to cases of prefrontal damage cited by Deacon (1997, pp.263–4): patients whose capacity for divergent thinking became impaired as a result of pre-frontal damage, apparently lose the ability to perform tasks involving 'short term memory, attention, suppressions of responses, and context sensitivity'.

Returning to the above objection, in view of the fact that the automatic link is only replacing the symbolic use of divergent thinking, it is not the case that we are losing divergent thinking entirely as a source of invariance: both divergent and interpretative uses would still be retained. In other words, the above objection does not apply to the interpretative use of divergent thinking, as part of information processing at the discourse level. In the case of the symbolic use, where it does apply, one does, indeed, lose one source of invariance; this loss, however, is compensated by the emergence of another source, as automatization entails persistent demands on the systems involved in information processing.

Thus, language use at the text level provides at least two sources of invariance. The first involves the use of divergent thinking – in keeping apart propositional and illocutionary levels and in deriving meaning in context (i.e. of specific or indirect speech acts on the basis

6 That there should be an interpretive use of divergent thinking entails that there is a distinct process whereby meaning in context is determined. The consistency checks that have been postulated at the sentence level would be dealing with the co-text, rather than the context. This said, inasmuch as the interpretative process can be retroactive, a certain amount of interaction between the two levels can be expected.

7 Arguably, prior to the linguistic encoding of illocutionary values, divergent thinking would also have been called upon in an interpretative capacity to determine the illocutionary force of lower level-structures.

of the initial illocutionary value); the second, the automatization of the decoding of illocutionary meaning from lower-level structures. A third source, which lies in a further automatization of information processing, involves making relevant contexts readily available, by encoding them as idealized contexts. Speech acts are not consistently associated with specific surface structures – and this, not even within the same language – so distinct SN conditions may be the most stable features they possess, which might act[8] as inputs to the recognition process.

At this point two questions arise:

(1) How do the demands made by the processing of information in the case of speech acts differ from those required at the sentence level? (The possibility of a distinct neural substrate for speech acts cannot be envisaged, unless this point is clarified.)
(2) What do SN conditions derive their invariance from?

Information processing at the speech act level

To the extent that both at the sentence and the speech act level one is dealing with the interpretation of symbolic relationships, there seems little reason for the processing demands to be different: indeed, both cases require 'a strategy of one rapid interpretation followed by another', where, in addition to 'rapid implementation', an ability is also necessary, which keeps 'previous operations from interfering with subsequent ones' (Deacon 1997, p.313). However, there are indications to the effect that things may be otherwise. We saw previously that interpreting an utterance in terms of a speech act does not entail rejecting its meaning at the propositional level. The attentional shift required to access a higher-order association does not result in propositional meaning being cancelled: interpreting 'I will get Octavio

8 Obviously, in the case of indirect speech acts SN conditions would be overridden by external contextual considerations. And when it comes to distinguishing between two speech acts of the same class, contextual intervention again would be necessary.

to apologize to you' as a promise does not entail the rejection of the lower-level interpretation, which concerns a future act of S; this interpretation merely recedes to the background. This awareness of parallel levels of interpretation appears to involve an ability also shared by the comprehension of jokes and stories, which requires that more than one network of (contextual) relationships be borne in mind simultaneously (Deacon 1997, p.313, and Kosslyn and Koenig 1995, p.245). In addition there is a further requirement, to do with processing speed: given the sheer amount of aspects to be attended to during the interpretation of a speech act, parallel processing – which is usually the norm, in any case – would be necessary. Now, there seems to be a general principle (invoked by Deacon 1997, p.315) to the effect that: when sensorimotor or cognitive operations tend to compete simultaneously for the same neural substrates, there is strong developmental selectional pressure to segregate the competing operations to counterpart structures in the opposite hemispheres to keep them from interfering with one another. In other words, it is the sameness of the processes operating at the propositional and the speech act level that would, owing to their sheer scale, lead to the co-opting of additional neural substrate at the speech act level.

One question this raises is: assuming that segregation makes sense in processing terms, why does it have to be between those two particular levels? A possible answer lies in the possibility that the speech act level marks a further level of symbolic recoding, one involving a different mode of coding. As will be recalled, it is the pathway corresponding to the process of using a sentence that would become linked to an illocutionary value, rather than the sentence itself. In other words, we could be talking here about the encoding of a motor representation, as opposed to that of an auditory or visual representation. In short, while the types of operation at work may remain the same, the possibility for a demarcation line does exist, which lies in a change in the nature of the inputs involved. A further source of rupture may lie in the nature of the pre-programmed interpretative hypothesis required for speed of interpretation: both at the sentence and speech act levels, such hypotheses (available from associative memory) consist of pre-programmed contexts. However, in the former case we are dealing with linguistic contexts, provided by pre-encoded sentences, while in the latter these contexts need not be exclusively of a linguistic nature.

SN conditions and invariance

Our next issue concerns invariance in respect of SN conditions. Since SN conditions are part of the conceptual structure of certain speech acts, as idealized in-built conditions ('idealized', in the sense of a full set of maximally distinctive conditions), it is not unreasonable to assume that, as contextual cues, they must have exhibited a high degree of invariance, in the sense of having a high level of occurrence over an extended period of time, as a full set of identical conditions; these conditions, moreover, would have been perceptible as such to interlocutors.

Now, while a high level of occurrence for speech acts poses no particular problem, how likely is it for naturally occurring factors to be invariant, both in terms of involving the full set of contextual conditions each time, and in terms of presenting identical conditions across situations? This would be implausible enough for conditions with perceptual attributes, let alone conditions which, for the most part, are mental. Thus, in the case of orders, one such condition is that S wants H to carry out an action A, another is that S believes H to be in a position to carry out A, yet another is that S does not think that H was going to do A of his own accord.

Underlying what precedes is the assumption that unless stimuli present a high level of invariance, they cannot give rise to invariant outputs; in other words, the role of the perceptual subsystems would merely be one of recording what is 'out there'.

Leaving aside, for the moment, the distinction between external and mental conditions, is it the case that invariance has to derive exclusively from external stimuli? This is counter to what is known of the way perceptual subsystems work. From the standpoint of a system whose main priority is processing efficiency, with speed as a major requirement, anticipation (backed up by revision-on-the-fly, should the need arise) is an essential strategy. Anticipation implies a built-in 'presumption of invariance' on the part of the system, with regard to both main and contextual stimuli. Such a system relies on ready-made solutions in the form of patterns of activation functioning as interpretative hypotheses, and on processes biased towards their preservation, in virtue of their capacity to generalize.

The extent to which invariance can arise from the processing system is evident in the types of process involved in identification tasks. Thus, although there is a matching process that compares perceptual representations of stimuli to stored patterns, an absolute match appears to be unnecessary for some of those patterns to be activated to one degree or another. In other words, even stimuli that are not familiar in every respect tend to be subsumed under the closest matches, rather than encoded as new memories (Kosslyn and Koenig, 1995, p.348). Furthermore, these patterns of activation themselves are prototypes resulting from an averaging process inherent in processing stimuli: assuming that processes in pattern activation subsystems can be expected to behave in a way similar to that of a computer simulation models with a recurrent network design,[9] this is how Kosslyn and Koenig (1995, p.336–7) envisage the formation of prototypes: each time a stimulus (corresponding to an exemplar) is stored in a pattern activation network, the weights on the connections are adjusted; these successive adjustments result in a pattern of weights which is a 'compromise between those formed for individual stimuli'. That categories 'are not formed by a process that delineates sets of defining features, and generates a list, which is then compared against the input' is consistent with Rosch's findings (Rosch 1978). Reliance on external invariance may be even less than expected if one considers that prototypical traits only effectively defines a category to the extent that they successfully contrast it with another category. Thus, prior to the matching test, not only are stimuli grouped by the preprocessing subsystems on the basis of nonaccidental properties and similar

9 A recurrent network design, as described by Kosslyn and Koenig (1995, pp.336–7), involves input, output and hidden units corresponding to neurons linked to one another with connections of varying strengths, or weights, which determine the level of activation or inhibition one unit has on another, to which it is connected.

features, but also in ways that have proved useful in the past.[10] This, in other words, presupposes a yet further set of internal criteria (Kosslyn and Koenig, 1995, p.76 and p.219).

Another aspect of this matching process that appears to exemplify this generalizing tendency of brain systems is found in the fact that inputs are matched to 'the largest possible units, each of which accounts for "more variance" in the input' (Kosslyn and Koenig 1995, p.224). This type of generalization, which concerns relationships between stimuli, may be due to what is known as the 'priming effect' (Kosslyn and Koenig 1995, p.378), whereby prior exposure to a set of stimuli will bias a system to respond to subsequent exposures to any one of these stimuli by activating the network of representations corresponding to the set in its entirety.

Mental conditions and invariance

Given that invariance need not be an original property of inputs, but can arise from processing, the next question is: can mental conditions acquire it?

Alternate scenarios

Inasmuch as mental conditions did make their way into associative memory as part of in-built contexts, they can be assumed to have acquired invariance through processing. The question is 'How?': since they are not external and do not possess sensory attributes, they cannot have gone through the system as external inputs. Could they, then, have done so as internal inputs? What lends a measure of plausibility to this idea is the possibility of a parallel with the visual system, where

10 According to Kosslyn and Koenig (1995, p.77), 'areas in the visual system receive feedback from higher areas [...] These feedback connections may have a role in "training" the preprocessing subsystem so that it can more easily encode visual characterizations that have been useful in the past'.

there is a fair amount of evidence that visual mental imagery recruits[11] the same processing systems used in visual perception (Kosslyn and Koenig 1995, pp.130–9); these mental images, based on representations stored in the visual pattern activation subsystem (a unimodal long-term memory), would engender representations in the visual buffer, as required by the task at hand. Assuming that this also applies to language, in the course of information processing, the sentence structure buffer[12] would, in the case of mental conditions, receive inputs from the corresponding pattern activation subsystems, instead of the preprocessing subsystems. This is how mental conditions could, conceivably, go through the system as internal inputs.

The obvious question this proposal prompts is: is it possible to have encoded patterns without initial sensory inputs in the first place? Moreover, what would that be in the case of mental conditions? Mental conditions are supposed to derive from 'Background knowledge'[13] (Searle 1983, 1995), which, as implicit knowledge, can only

11 For further details regarding the use of the same subsystems by visual imagery and visual perception, cf. Perrett et al. 1985.

12 A sentence structure buffer is 'a short-term memory structure that stores syntactic and semantic information as the meaning of a sentence is built up over time' (Kosslyn and Koenig 1995, p.389). This information (conjointly with the prosodic information it also stores) places strong constraints on how words may be connected, thereby monitoring inconsistencies at the sentence level (Kosslyn and Koenig, 1995, pp.242–3).

13 Searle (1983, pp.142–3) defines the Background as 'a set of nonrepresentational mental capacities that enables all representations to take place. Intentional states only have the conditions of satisfaction that they do, and thus only are the states that they are, against a Background of abilities that are not themselves intentional states. In order that I can now have the intentional state that I do I must have certain kinds of know-how: I must know how things are and I must know how to do things, but the kinds of know how in question are not, in these cases, forms of know-that.' The sense in which the Background is nonrepresentatonal is that, like 'implicit memories', it is not accessible to introspection or available for other purposes, but are only activated in performing skills. Thus 'For me the hardness of tables manifests itself in the fact that I know how to sit at a table, I can write on a table, I put stacks of books on tables [...] And as I do each of these things, I do not, in addition, think unconsciously to myself "it offers resistance to touch.' Even though what Searle says of the Background appears to apply to Kosslyn and Koenig's implicit memories, his notion of

be activated in performing skills, and is not, in principle, accessible to other systems. Thus, people's knowledge of mental conditions would only manifest itself in their ability to make and handle speech acts, and could not be available for other purposes (e.g. as a basis for reasoning). In terms of the memory subsystems, the implication is that they cannot engender a representation in the sentence structure buffer, and a good indication that that may be the case is that, while handling speech acts, one would not normally be aware of them. As Kosslyn and Koenig (1995, p.388) put it, 'If we are not aware of it, the information is not in short-term memory.' So, unless there is a process whereby implicit knowledge can make its way into the pattern activation subsystem (and from there into the sentence structure buffer, where it becomes available for other purposes, before being encoded in associative memory), another alternative should be envisaged.

The one I propose to consider is that mental conditions made their way into associative memory through the explicit memory formation process: on this view, the pattern activation subsystems and the sentence structure buffer would also be involved, but part of the input would be external.

But before going into the detail of this scenario, what indications are there that we actually possess explicit, as well as implicit, knowledge of speech acts, including SN conditions? That we possess implicit knowledge of speech acts and of their contextual factors is not in doubt, since we know how to perform, identify and respond to them

representation is entirely derived from Speech Act Theory. Thus, to Searle 'the sense of "represent" in which a belief represents its conditions of satisfaction is the same sense in which a statement represents its condition of satisfaction. To say that a belief is a representation is simply to say that it has a propositional content and a psychological mode, that its psychological mode determines a direction of fit, of its propositional content, in a way that all these notions – propositional content, direction of fit, etc. – are explained by the theory of Speech Acts' (1983, p.11–13). By contrast, Kosslyn and Koenig (1995, p.27), use 'representation' in the sense of 'something that stands for something else.' Thus, 'If we think about a network that recognized objects [...], the patterns of weights in the network will be a representation of different objects.' For a different notion of representation, see also Searle 1983.

appropriately. This, however, does not entail that we have the equivalent explicit knowledge. Implicit knowledge, as will be recalled, can only manifest itself in the performance of skills (Searle would call it 'preintentional'). It is not available to systems other than those mediating these skills and cannot be used for other purposes (e.g., as a basis for reasoning). In other words, this implicit knowledge cannot be expected to give rise to representations[14] of the context, in the sense of manifestations of the corresponding knowledge that were separate from skill performing.

The fact still remains that we appear to have access to contextual knowledge about speech acts: besides producing, categorizing and responding to them automatically and appropriately, we also challenge them, as shown by responses such as 'I was going to do it anyway!', 'I don't take orders from you!', 'You are not serious!', when compliance with an order seems out of the question. In all these instances there is reference to a distinct mental SN condition – or rather to the alleged fact that it did not obtain – for the purpose of countermanding an order.

Further support for this hypothesis lies in the existence of associative memory. Among other things, this memory stores the results of explicit memory encoding, which include the contexts in which stimuli occur. That SN conditions, in general, could have made their way into associative memory via explicit memory encoding is by no means implausible, as they arise from different modalities (e.g. visual, auditory), and associative memory is amodal, in the sense that it is 'where outputs from different perceptual processes come together' (Kosslyn and Koenig 1995, p.83). Unlike associative memory, the perceptual subsystems (which, incidentally, mediate the stimulus-response connection system, among other things) are modality specific.

On the basis of what precedes, representations of the context could well have formed as part of explicit memory encoding, with the context of utterance providing the association between the utterance of a lower-level structure and the lower-level structure an illocutionary value.

14 See note 13.

This caters for the question of how implicit knowledge could have made its way into the 'mainstream' processing system thereby acquiring invariance. The answer we have arrived at is, in a nutshell, that the question does not arise, since there are two types of memory formation, and that what is inside associative memory made its way there via explicit memory formation. What this does not tell us, though, is how mental conditions could have gone through this process, since they do not possess sensory attributes.

Mental conditions and sensory correlates

As a preliminary to addressing this question, let us consider the following objection. Our concern – it could be pointed out – is misplaced, in view of what really goes on when we interpret a speech act. Take, for example, an order: if we decide that the sincerity condition does not obtain, nothing further need follow. On the other hand, if we believe this condition to be fulfilled (and are satisfied that S is in the proper position of authority), then, in virtue of our background knowledge (cf. Searle 1979, 1983), we will necessarily believe that S cannot sincerely want H to do A unless he also has all the appropriate mental states (i.e. believes that A is feasible, that H is in a position to bring about A, etc.) It is unclear what further need would be served, if H were to attempt to find out whether S actually possesses those mental states. To summarize this objection: once the sincerity condition is deemed to have been satisfied, the other mental conditions can be taken for granted. In other words, it seems pointless to ask whether they are amenable to verification, or whether they are perceptible.

The above objection is not relevant to our concern, which is with how mental conditions make their way into the conceptual structure of certain speech acts, and not with how H goes about ascertaining that they obtain. Perhaps it would be helpful to restate the problem. Given that:

(a) SN conditions, including mental ones, appear to be stored in associative memory, as in-built context;

(b) only stimuli, with sensory attributes, can serve as inputs to the perceptual subsystems that constitute the starting point of the explicit memory formation process, how did mental conditions make their way through this process (thereby acquiring their invariance)?

On the basis of what was said earlier, the answer that suggests itself is that they did so via sensory inputs. Although mental states are not perceptible as such, they can be correlated to stimuli susceptible of being interpreted accordingly, especially if relevant contextual cues are also present, which facilitates a convergence on that interpretation. Thus what goes through the perceptual subsystems could be the utterance's intonation pattern, S's tone of voice, the way he behaves, as possible indications of his intention. The occurrence of those stimuli would activate an association between perceptual representations and background knowledge about the relevant mental states.

The question this prompts is: is it conceivable that each mental condition will have a distinct perceptual correlate? Consider, for example, the self-referential causality condition (which, in the case of orders, stipulates that the action required of H has to be caused by the order, rather than by other reasons), or the condition that S must believe that H can do A, or this further condition that S must believe that H was not going to do A, of his own accord. In all these cases, it is unclear what would constitute an appropriate sensory correlate. The only condition that appears to possess such a correlate is the sincerity condition. Alternatively, since S cannot possess the mental state corresponding to the sincerity condition unless she also possesses the subsidiary mental states, one may decide to view this correlate as a collective one, corresponding to the whole set of mental conditions.

Assuming that we are dealing with collective sensory correlates, a host of questions arise:

(a) What processes would be responsible for sorting these correlates from accidental stimuli?
(b) How do these collective sensory correlates come to be associated with individual conditions?

(c) How do such conditions make their way into associative memory?
(d) How do these individual conditions acquire invariance?

The answer to (a) appears to lie in the operation of the preprocessing subsystem. As will be recalled, this subsystem categorizes stimuli on the basis of past usefulness (Kosslyn and Koenig 1995, p.219). The way this is implemented is through reciprocal connections between the preprocessing subsystem and higher-level ones: these connections allow the former to receive feedback from the latter, (since 'in general an area that sends information to another also receives information from it.') and positive feedback (to the effect that the identification task was successful) will reinforce the encoding of corresponding properties in the preprocessing subsystem. (The fact that the sensory correlates of mental conditions in question occur alongside stimuli corresponding to other conditions would also facilitate this process.)

Once marked as relevant by the preprocessing system, these sensory correlates would need to be associated with individual conditions (= (b)). This can be achieved via reciprocal connections with higher-order systems, in the event, a conceptual memory which holds our knowledge about the above conditions. Invoking such a memory is not without prompting a set of questions for which no immediate answers appear to be available, at least not on the basis of the assumptions introduces so far. Thus:

- How does the knowledge it stores differ from the implicit knowledge of speech acts?
- What kinds of inputs does it require?
- In what way does its contents differ from that of a long-term memory
- Which is not conceptual? Alternatively, what do concepts and perceptual categories have in common and what separate them?
- Would a conceptual memory be distinct from associative memory?
- Would such a memory be prelinguistic?

If such a conceptual memory can be assumed, the way the encoding in associative memory (= (c)) of the above conditions would come about as a result of tasks that single them out, thereby increasing their likelihood of being remembered. This 'depth of processing effect' (Kosslyn and Koenig 1995, pp.375–6), is characteristic of the process whereby relevant information is activated in the course of interpretation. It is also associated with tasks involving explicit reference to individual conditions, for the purpose of invalidating a prior speech act (as we have seen above), or in order to reassure H of the validity of one's speech act (as in 'I have the firm intention of fulfilling my promise!'). Awareness of subsidiary mental conditions would arise, as a matter of course, each time the outcome of one's speech acts is a cause for concern; thus wondering about the feasibility of an action A, prior to making a request that it should be carried out, would give rise to a representation of the corresponding condition[15] in the sentence buffer. Once in the sentence structure buffer, such condition can, in principle, make its way into associative memory, as reciprocal connections also appear to exist between the two types of structure, which allow for interplay between short and long-term information, as part of the working memory system.[16]

All the above tasks give rise to depth of processing, which functions to reinforce explicit memory encoding, since more than one type of information and connection would be activated.

Turning finally to (d), in light of (b), it is unclear how individual mental conditions can acquire invariance via collective sensory

15 Awareness of a piece of information is taken to be a sign that it is in short-term memory (Kosslyn and Koenig, 1995, p.388).

16 According to Kosslyn and Koenig (1995, pp.387–8), memory representations can be held in short or long-term memory structures. Structures of the former kind include the auditory, the visual, and the sentence structure buffer; those of the latter kind, the pattern activation subsystems, motor memory (which are modality specific) and associative memory (which is amodal). Tasks involving reasoning require interplay between short and long-term memory representations, as well as some form of temporary storage for them. This led to the postulation of a working memory system consisting of activated information in short-term memory and decision-making processes in the frontal lobe, which 'selectively activate information in long-term memory and "swap" information in and out of short-term memory, as needed to perform a task'.

correlates, since they are yet to be differentiated at that stage. A more satisfactory hypothesis would have the same apparatus that imparts invariance to perceptual categories play a key role in concept formation: mental conditions would then derive their invariance from the relevant concepts.

In this scenario invariance would be imparted to all SN conditions by internal mechanisms, but the categorizing process would take internal as well as external inputs. Such a hypothesis requires a theory of concept formation that fits in with the above picture and provides a clear answer to the questions raised under (b).

This is where we will need to turn to Edelman's extended Theory of Neuronal Group Selection (TNGS), which will be presented in Chapter 4. Chapter 3, however, will have more to say about the contents of the memories at hand, and what inputs they require.

Summary

The eligibility of certain SN conditions for encoding depends on whether they can present the appropriate level of invariance. Concerning the source of invariance, we ruled out the possibility that it can only be external, in favour of the hypothesis that it is to a great extent imparted by the processing system. At this stage, the question arose as to whether mental conditions, like perceptible ones, could acquire invariance in this manner. The first proposal we put forward is that unless there is a process whereby implicit knowledge, such as underlies skills, could make its way into the relevant short-term memory (where it could be processed as sensory stimuli, acquire invariance and become encoded in associative memory as explicit knowledge), the alternative would have to be that it made its way there in a more direct fashion, via the explicit memory formation process. At this point the same worry regarding mental conditions returns: how can they go through the explicit memory formation process, unless they possess sensory attributes? The possibility that then suggested itself was that they would have to have sensory correlates instead, or rather share the same correlate. The idea of a collective sensory correlate, however, does not cater for the problem of how individual mental conditions can acquire invariance. A more satisfactory solution, we ended by proposing, lies in having invariance derive from a conceptual source, which is a distinct possibility if the categorizing mechanisms can be assumed to apply to concept formation. This possibility is one allowed by Edelman's TNGS.

Underlying means

To bring this chapter to a close, I propose to go over the main underlying means language could have co-opted. This co-option, as will be recalled, is deemed necessary if language processing is to be maximally efficient in coping with new demands at the text level, in the event those connected with the interpretation of illocutionary acts and the obligations they impose.

In terms of organization, what is first needed is a list of features that may plausibly subserve the requirement of processing efficiency, to which all processing systems are subject. With this list on hand, we can then ask to what extent the features it contains meet the demands of language processing at the text level, and what further means are required.

Features which, among other things, contribute to processing efficiency fall under two categories, depending on whether their main (or only) function is to ensure speed of processing or accuracy. (This way of categorizing them does not entail that the same feature, or different aspect of it, cannot mediate both functions.)

Features subserving processing speed

Features subserving processing speed include:

(a) A tendency towards automatization, best exemplified by the stimulus-response connection system,[17] which links perceptual

17 The stimulus-connection subsystem is implemented in the striatum. The latter, which is part of the basal ganglia, directly connects perceptual inputs and motor

and motor responses directly, thereby enabling the processing systems to bypass intermediate stages, once the relevant connections have become well-established.

(b) An ability to anticipate an outcome (as opposed to working it out). Such a tendency is mediated by the perceptual processing subsystems' capacity for generalization. Generalization gives rise to encoded activation patterns corresponding to properties of classes of stimuli. Having such patterns enables processing systems to anticipate what kind of stimulus they encounter in the course of identification tasks. The ability to anticipate an outcome is also based on the fact that activation patterns corresponding to objects encoded via the explicit memory encoding system also include features of the context in which these objects occur.

(c) An ability to reach an outcome on the basis of incomplete evidence. This ability, one of converging on a solution on the basis of a set of cues – which, taken individually, would be insufficient to constrain an outcome – is dependent on the existence of systems operating in parallel and having, to varying degrees, sensitivities to the same range of inputs. Each of these systems would also be maximally sensitive to one particular type of input. A typical example (cited by Kosslyn and Koenig 1995, pp.47–8) is that of three types of cells (or cones) the eye contains, which register colours: although each type responds maximally to a specific wavelength, corresponding to red, green and blue, respectively, it also responds – albeit less vigorously – to a range of shorter wavelengths. As a result, we can perceive colours such as brown or orange, which provide a better indication of a particular object. This property is known as 'coarse coding'. In addition to producing responses that overlap, these systems are subject to 'constraint satisfaction' (Kosslyn and

programs ('in part' because [it] receives information from cortical perceptual areas and indirectly (in part via the thalamus) sends fibers to the premotor cortex and the supplementary motor cortex.' (Kosslyn and Koenig, 1995, p.347)). See also Miskin 1978, Cohen and Squire 1980, Mishkin, Malamut and Bachevalier 1984, Mishkin and Appenzeller 1987.

Koenig 1995, pp.46–9), a requirement to the effect that all the constraints (concerning the various types of input) should be satisfied at the same time. If individually each cue only weakly constrains an outcome, as a set to be taken into account simultaneously, they provide a very strong constraint.

Features underlying accuracy

This second category of features, whose function offsets that of the first category, includes:

(a) The explicit memory formation process, which involves associating the information to be remembered with its context of occurrence (Kosslyn and Koenig 1995, p.317). Having activation patterns encoded with contextual information not only contributes to speed of processing but also makes for a more accurate identification of stimuli.

(b) Coarse coding, which also caters for accuracy: the solution arrived at by coarse coded systems operating in parallel also happens to be more accurate, since the information they produced overlap. Coarse coding is 'the method of using the degree of overlap in responses from units (or entire networks) that have different sensitivities to specify precise values' (Kosslyn and Koenig 1995, p.47–8). Consider this example from visual cognition, cited by these authors: although we perceive colours and orientation independently, many neurons in the visual area known as V4[18] respond to both colours and orientation[19] of stimuli to varying degrees. Being sensitive to different

18 V4: The reference is to van Essen and Maunsell's (1983) diagram of visual areas (*Trends in Neurosciences* 6, pp.370–375).

19 Neurons in visual area MT, which also respond to motion, provide information to subsystems that distinguish shapes from background as well as to those responsible for tracking moving objects (Kosslyn and Koenig 1995, p.45). This is a consequence of weak modularity, another feature of processing subsystems, whereby the latter share components at a finer level of analysis, instead of being totally independent, as was once proposed by Fodor (1983).

degrees, these neurons produce an overlap of information, which allows for the emergence of more precise values. Thus, taken together, the multiple cues provided by a set of neurons constitute 'a better indicator of edges than any single cue'. This use of an information overlap also balances out the disadvantage of concurrent processing, which produces a lot of degraded information.

(c) Reciprocal connections: these mediate the exchange of information between lower and higher-order systems (e.g. perceptual preprocessing and processing subsystems, the working memory system, which includes the sentence structure buffer, on the one hand, and long-term memories, on the other). Such connections make it possible for current categorizations to be selected on the basis of their past usefulness, and, more generally, for conceptual knowledge to be brought to bear on the operation of lower-order systems.

While by and large subject to the requirement of processing efficiency, the processing of speech acts also has demands of its own, arising from the following facts:

(1) Illocutionary meanings are expressed by lower-level structures already associated with lower-level meanings.
(2) These lower-level meanings cannot be ruled out, but have to be retained in working memory and integrated to the corresponding higher-level meanings, to which they stand in a hierarchical relation. (Propositional meaning, as will be recalled, provide illocutionary acts with their propositional contents.)
(3) The identification of higher-level meanings also requires quick access to sufficient contextual features, in the event, SN conditions.

Given that the identification of speech acts requires a considerable amount of inferencing, straightforward automatization has to be ruled out. On the other hand, the ability to anticipate an outcome on the basis of encoded patterns of activation, with what it entails in terms of supporting systems and structures, is a definite candidate for co-option, as it would cater for the need to provide quick access to SN conditions. As a matter of course, the formation of such patterns depends on the operation of the explicit memory formation system, the process of category formation inherent in the perceptual processing subsystems, the 'information exchange' system mediated by reciprocal connections between lower- and higher-order systems.

Another good candidate for co-option is the ability to reach an outcome on the basis of incomplete evidence. Such an ability comes in handy at the text level, where there is a wide range of cues, which, individually, are insufficient to lead to a solution. (Examples of such cues include linguistic features of utterances, the context of utterance, contextual assumptions, the way speakers make speech acts, etc.)

Conjointly these two abilities cater for (3). (1) and (2) require that we bring in divergent thinking, a mnemonic strategy that enables the interpretative process to find a shortcut between lower-level structures and higher-level meanings, while retaining in memory lower-level meanings. Divergent thinking also makes it possible for lower and higher-level meanings to be placed in a hierarchical relation. Shortcuts thus created provide the basis for further symbolic recoding at the text level.

In light of the additional burden imposed on reasoning abilities at the text level, further co-option seems likely, but this time involving additional neural substrate. Reasoning abilities are believed to be more effective in the right hemisphere (for right-handed people), especially when it comes to handling multiple contexts simultaneously. Right hemisphere patients, indeed, appear to experience difficulty in understanding jokes and metaphors, which involve the ability to keep track of more than one context at the same time (Kosslyn and Koenig, 1995, pp. 243–6). Now, if this is the case, it has been suggested that the right hemisphere might be co-opted as additional neural substrate for language processing at the text level. As mentioned earlier, there appears to be a general principle to the effect that when sensorimotor or cognitive operations tend to compete for the same neural substrates, there is strong developmental selectional pressure to segregate the competing operations to counterpart structures in the opposite hemispheres (Deacon, 1997, p.315). Such a segregation prevents simultaneous operations from interfering with one another, thereby also contributing to processing efficiency.

Chapter 3

Argumentational meaning

The emergence of argumentational meaning ('argumentatif' in AWL's terminology), would be another instance of language adaptation to processing efficiency, one also involving an interpretative shortcut between lower-level structures and higher-level meanings. This particular case, however, presents a problem of its own, which concerns the source of these higher-level meanings. In terms of co-optable underlying means, divergent thinking is the obvious answer to the first problem. In addressing the second we will need to turn to Antonio Damasio's model of decision making.

But first, what is argumentational meaning? Argumentation, as an aspect of meaning, has to do with constraints on continuation, as attested by the existence of preferred continuations. Consider the following examples, found in Anscombre (1995a):

(1) a Pierre est riche.
Pierre is wealthy.
b Il peut s'offrir n'importe quoi.
He can afford anything he fancies.
c Il est donc avare.
He is therefore stingy.

(2) a Marie est belle.
Marie is beautiful.
b Elle séduit tous les hommes.
All men all fall for her.
c Elle est donc coquette.
She is therefore a flirt.

In both sets of examples (a) is compatible with either (b) or (c) as a continuation; (b), however, appears to be the preferred option: in a context free situation, it is, par excellence, the conclusion one would draw from (a). (c) cannot achieve this level of naturalness unless certain contextual assumptions are brought to bear: (1) thus requires an assumptions to the effect that wealthy people tend to be tight-fisted; (2), an assumption associating beauty and flirtatiousness. The existence of such preferred continuations constitutes, in AWL's view, a clear indication of a close conceptual link between the two contents involved: between wealth and having the buying power wealth confers, or between beauty and the capacity to attract people. Such continuations Anscombre (1995b, p.57) views as mere explicitations of some feature inherent in the lexical items at hand: 'le second membre ne fait qu'expliciter (que "déployer") ce qui est déjà potentiellement contenu dans le premier' ('the consequent merely renders explicit ("unfolds") what is already latent in the antecedent'). The same idea is also expounded in Bruxelles, Ducrot and Raccah (1995).

In keeping its non-objectivist[1] conception of lexical meaning, the deep structure of lexical items, as construed by AWL, includes intrinsic topoï ('topoï intrinsèques'), for which the following may serve as an informal definition: 'La notion de topos intrinsèque correspond à l'idée qu'on ne peut parler de richesse sans évoquer le pouvoir d'achat, de beauté sans évoquer l'attirance qu'elle exerce, d'avarice sans mettre en scène la répugnance à donner qu'elle implique' (Anscombre 1995b, p.58). ('The notion of intrinsic topos serves to convey how the sheer mention of wealth will bring to mind buying power, that of beauty, attractiveness, that of stinginess, behaviour denoting a clear loathing for parting with money.') Intrinsic topoï, AWL insists, should not be confused with extrinsic topoï ('topoï extrinsèques'), an example of which is *[the more money someone has,*

1 'l'occurrence d'un mot ne réfère pas un objet (en tant que faisceau de constants), mais induit une certaine vision de la situation (en tant que faisceau de topoï)' (The occurrence of a given word does not constitute a reference to an object (as a bundle of invariant traits), but implies a certain take on the situation (construed as a bundle of topoï)) (Anscombre 1995b, p.51).

the more stingy he is], an assumption evoked earlier in connection with (a) and (c), in (1); a further example, provided by (a) and (c) in (2), is *[the more beautiful someone is, the more flirtatious she is]*. Unlike intrinsic topoï, which form an integral part of the meaning of lexical items, extrinsic topoï are thus contextual assumptions, whose function is to validate inferences, in the event, the process whereby certain conclusions are drawn from certain premises. This distinction allows AWL to account for a number of linguistic phenomena. Consider the two sets of examples below, which involve *mais* (but):

(3a) Pierre a cherché une solution, mais il n'a pas trouvé.
Pierre looked for a solution, but he failed to come up with one.
(4a) Il y a un problème, mais il est facile à résoudre.
There is a problem, but it can be sorted out.
(3b) Pierre a cherché une solution, mais il l'a trouvée.
Pierre looked for a solution, but he found it.
(4b) Il y a un problème, mais il est difficile à résoudre.
There is a problem, but no solution seems to be forthcoming.

Just as the naturalness of the sequencing in the (a) set can be explained in terms of a conceptual link between 'chercher' (looking for something) and 'trouver' (finding it) on the one hand, and between 'être un problème' (being a problem) and 'être difficile à résoudre' (not being amenable to an immediate solution), on the other, so can the oddity of the sequencing in the (b) set be put down to the cancellation of that link brought about by 'mais' and the proposition it introduces.

Another way of highlighting this preferential link is by means *avec succès/ sans succès*, another criterion proposed by Anscombre. Consider:

(5) Il a cherché avec succès.
He was successful in his search.
(6) Il a cherché sans succcès.
He was unsuccessful in his search.

In a context free situation (5) and (6) can only be taken to mean (5') and (6'), respectively, thus confirming the existence of a preferential link between 'chercher' and 'trouver':

(5') He found what he was looking for.
(6') He did not find what he was looking for.

Inasmuch as the above data are clearly theory-bound, and I am clearly not utilizing AWL as a general framework, a natural worry is how much of what makes up the specificity of this data would remain once the theoretical trappings have been removed.

While I do not know to what extent there can be an observational language which is unbiased by parameters and categories arising from identifiable theoretical concerns,[2] the above worry can be somewhat mitigated. Indeed, the data under consideration are amenable to an analysis from a different perspective without incurring any substantial loss in terms of features which are relevant to our purposes. The perspective I have in mind is Searle's theory of the Background (Searle 1983, 1995), which is part of his more general theory of Intentionality. Thus, the type of content exemplified by extrinsic topoï bears a clear resemblance to Searle's cultural and local background assumptions, as for the situations corresponding to the consequents of intrinsic topoï, they evoke abilities, stances and practices that are part of our knowledge of the world, that is, of how things are, and how to engage with them.

Searle's framework may not make any provision for intrinsic topoï as an encoded aspect of meaning, but it provides them, nonetheless, with a sufficient level of conceptual specificity to enable us to view argumentational meaning in terms of underlying processes. Deferring a presentation of the Background to a more appropriate point, this is what viewing topoï in intentional terms looks like: where AWL sees preferred continuations as mere explicitations of some aspect or property of entities being referred to, in terms of the Background, such continuations constitute representations of stereo-

2 This linked to the assumption that things, if perceived at all, have to be so under some aspect or another.

typical situations featuring abilities, stances and practices associated with their handling. Thus (1) exemplifies a linkage between a property (being wealthy) with the kind of practice those blessed with it, can engage in; (2), one between a property of another kind (beauty) and a certain ability (that of attracting people) inherent in it; (4) is a case of an object (a problem) being associated with certain expectations regarding the amount of effort required in dealing with it. (3), though arguably not about a stance, an ability or a practice, is nonetheless a matter of ideology, rather that anything entirely predictable on the basis of some property of the process under consideration: while finding may be a possible outcome of seeking, not all instances of seeking result in success. In other words, the intrinsic topos in this case would also be construable in terms of the Background, albeit the local Background, where cultural assumptions are to be found, rather than the deep Background.

Argumentational meaning, as just described, would have emerged as the end-result of a complex process whereby lower-level entities already conventionally associated with a lower-level meaning have become – if not quite re-encoded as such – at least strongly connected with higher-level contents pertaining to background knowledge. It is in virtue of these contents that the occurrence of those lower-level entities will constrain certain types of continuation.

A more complete scenario of the emergence of argumentational meaning would have to address the following issues:

(a) How must those higher-level contents be construed, if they are to do the validating job just described?
(b) On what grounds are they to be thus construed?
(c) What kind of linguistic activity would have facilitated the linkage between a lower-level structure already conventionally associated with a lower-level meaning and a higher-level content?

Of these, (a) and (b) correspond to what was referred to earlier as the 'knowledge source problem', as they are about the way our knowledge and experience of the world are categorized, and how they come to be that way. (c) is the interpretative problem.

Concerning the knowledge source problem, I propose that:

- Those higher-level contents should be construed as complex categories, consisting of categories of entities or situations linked to other categories of situations.
- These complex categories came about very much in the same way as categories of situations, categories of response options and categories of outcome do, according to Damasio (1994)'s model of decision making. As a matter of fact, they can arguably be regarded as part of an extended version of Damasio's store of categorized factual knowledge. The assumption behind this is that the communicative situation in which argumentational meanings occur is, in some key respects, similar to the decision-making situation as described by Damasio.

Regarding the interpretative problem, my assumptions are as follows:

- The linguistic activity that would have facilitated the further encoding of lower level entities with constraints on continuation is an interpretative process, one involved in interpreting lower-level structures as expressing a certain type of higher-level meaning, in the event, S's desire that H should correctly identify or anticipate the upcoming text. (Such a higher-level meaning would typically be exemplified in acts of argumentation.)
- Divergent thinking underlies the discovery of the interpretative shortcut, thereby paving the way for the encoding of constraints on continuation.

In terms of organization, addressing the knowledge source problem involves:

(1) Explaining on what grounds the solutions to this problem should be drawn from Damasio.
(2) Providing a presentation of his decision-making apparatus.
(3) Extending the repository of categorized knowledge he postulates to include – alongside a knowledge based on individual experience – another type of knowledge, derived, this time, from

shared experience. This is where a more thorough presentation of Searle's theory of Background will be required.

Addressing the interpretative problem entails specifying:

(1) What type of interpretative process we have in mind, in terms of the type of communicative act it serves to interpret. Such an act, in my view, would have to involve categorization and reference to complex categories, if the corresponding interpretative process is to lead to an eventual encoding of anything remotely like complex categories.
(2) In what respects this interpretative process is similar to a decision-making process, and where it could exhibit a further level of automatization.
(3) What role divergent thinking has in the creation of an interpretative shortcut.

The knowledge source problem and decision making

Why Damasio's decision-making apparatus?

Our proposed answer to the knowledge source problem was that the higher-level contents in question should be construed in terms of complex categories, similar to those found in Damasio's repository of factual knowledge, which is available to decision-making. What makes Damasio's hypothesis so appealing? For one thing, argumentational meaning does not occur in a vacuum, but as part of a certain type of communicative situation, with a fairly specific communicative goal. On the basis of the above examples, this communicative goal is, by and large, one of getting H to identify a certain continuation (often construable as a conclusion) on the basis of an initial situation. Either way, on the basis of the overall situation, we appear to be dealing with a case of decision making – albeit one complicated by the fact that it is an object of representation (involving symbolic means) and constitutes an attempt to induce a similar process in H. The extent to which this process conforms to a rational model of decision making is something we will be in a better position to ascertain when the details of both Damasio's apparatus and of the relevant communicative situation have been clarified.

The second reason for looking to Damasio is that inasmuch as decision making at the social and linguistic level has an impact on survival, it can be expected to avail itself of supporting structures that already exist at the neurobiological level. As a matter of fact, in some important respects, decision making as inherent in language comes closer to those unconscious models more directly rooted in the neurobiology than to the rational[3] model, that is, one exhibiting the following parameters:

The decider has knowledge (a) about the situation which calls for a decision, (b) about different options of action (responses) and (c) about the consequences of each of these options (outcomes) immediately and at future epochs (Damasio, 1994, p.166).

In addition to this knowledge consisting of complex categories involving types of situation paired with types of response and types of outcome, the decider would have to have a strategy – logical or otherwise – to select the most advantageous outcome within a given time frame.

Finally, constraints on continuation, whose function is to ensure textual cohesion, may be seen as selection criteria arising from a highly automated decision-making process. Why this should be of particular significance has to do with the connection Damasio (1994, pp.198–200) makes between decision making and 'the critical problem of creating order out of parallel spatial displays, a problem [...] which arises because the brain's design only permits, at any one time, a limited amount of conscious mental output and movement output.'

3 For a thorough treatment of rationality in action see Searle 2001.

Antonio Damasio's decision-making apparatus

According to Damasio, biological processes culminating in a response selection fall into three categories. The first is made up of processes that are, for the most part, unconscious. Thus, when neurons in the hypothalamus detect a drop in the blood sugar level, this automatically triggers a hunger state. There is no awareness of a situation calling for action, no knowledge of response options or of a reasoning process leading to a response selection that might be brought to consciousness; all there is is a:

> physiological 'know-how' as inscribed in the dispositional representations[4] of the hypothalamus; and inscribed in a neural circuit, there is a 'strategy' to select a response consisting of instituting a hunger state, which will eventually drive you to eat (Damasio 1994, p. 166).

The second category contains processes that are in part conscious. A typical example is a situation where one happens to be in danger of being hit by a falling object. As in the previous case, the response (to duck or catch) will automatically be triggered. Clear differences, however, separate the two cases: only in the latter is there awareness of what type of situation one is dealing with; furthermore, the 'know-how', instead of innate, is acquired from experience, in the event, via exposure to falling objects and consequences attendant upon them. As a result of repeated experience of such scenarios, a strong connection is established between the provoking stimulus and the most advantageous response. No more than the previous case, does this one involve a strategy for response selection, or any response selection at all, for that matter. All there is is an automatic response

4 Alongside this type of dispositional representation, which is innate, there is another one, which is acquired, embodying 'knowledge pertaining to how certain types of situation usually have been paired with certain emotional responses in your individual experience.' Damasio 1994, p.137. Dispositional representations, as defined by Damasio (1994, p.102) 'exist as potential patterns of neuron activity in small ensembles of neurons [or "convergence zones"]; that is, they consist of a set of neuron firing dispositions within the ensemble.' As such they constitute our full repository of knowledge, both innate and acquired.

implemented through the strong link between stimulus and response being activated.

The third category comes closest to what may be regarded as a model of rational decision making, with a strong reliance on logical thinking, and much deliberation in the way each stage in the process is carried out. This category includes, on the one hand, how one goes about deciding whom to trust, whether to run a marathon, whether to hand in one's resignation; and on the other, the reasoning process that accompanies tasks such as solving an engineering problem, designing a house, or writing a book.

What characterizes this third category is the sheer number of options available at each stage, especially where consequences are concerned, whose advantages and disadvantages have ramifications in the future as well as the present. In view of this level of complexity, the task of reaching reliable predictions, especially if carried out consciously, is bound to be extremely time consuming as well as arduous and fraught with uncertainty in respect of its outcome. Nonetheless, as Damasio (1994, p.172–3) points out: 'our brains can often decide well, in seconds, or minutes, depending on the time frame we set as appropriate for the goal we want to achieve.' What underlies this possibility? This is where Damasio's somatic marker hypothesis comes in. As a first approximation, somatic markers may be thought of as arising from awareness of certain types of emotions,[5] those that in one's own experience are triggered by certain types of entity or situation:

> Those emotions and feelings have been connected, by learning, to predicted future outcomes of certain scenarios. When a negative somatic marker is juxtaposed to a particular future outcome, the combination functions as an

5 Emotion is 'the combination of a mental evaluative process, simple or complex, with dispositional responses to that process, mostly toward the body proper, resulting in an emotional body state, but also toward the brain itself (neurotransmitter in the brain stem), resulting in additional mental changes…' (Damasio, 1994, p.139) Damasio distinguishes between primary and secondary emotions. The former are innate, the latter acquired through experience. For further details, see Damasio, 1994, Chapter 7.

alarm bell. When a positive somatic marker is juxtaposed instead, it becomes a beacon of incentive (Damasio 1994, p.174).

A somatic marker is acquired when selecting a given option leads to an outcome that causes a body state (e.g. pain or pleasure), thereby giving rise to a dispositional representation for this connection. By virtue of such a representation, subsequent exposures to the same option or the thought of the same outcome will trigger the same body state (which functions as a deterrent or an incentive, as the case may be). By highlighting certain options as desirable or deleterious, somatic markers narrow down the range of alternatives requiring consideration, and, in so doing, considerably simplify the selection process.

Our next concern is with the origin of somatic markers. Somatic markers apparently arise within a system as a result of a learning process, whereby categories of entities or events are linked to pleasant or unpleasant body states. This process is based on the mechanisms of primary emotions followed by those of secondary emotions.

Primary emotions, which are innate, are automatically released when perceived stimuli present certain types of features, such as:

size (as in large animals), […] type of motion (as in reptiles), […]; certain configurations of body state (as in the pain felt during a heart attack). Such features, individually or conjunctively, would be processed and then detected by a component of the brain's limbic system, say, the amygdala [whose] neuron nuclei possess a dispositional representation which triggers the enactment of the body state characteristic of the emotion fear, and alters cognitive processing in a manner that fits the state of fear (Damasio 1994, p.180).

In contrast with primary emotions, which are wired-in at birth and connected with types of features of stimuli, secondary emotions would be acquired as we grow up (but also throughout life), and paired with entities and situations specific to individual experience. Thus most people hiking at night on a trail in the Back Country would probably have a fright if something were suddenly to loom large at their side; by contrast, how each person responds to minimalist interiors or concrete music is very much a matter of personal experience. As an instance of pairing of categories of stimuli and somatic states, secondary emotions, not surprisingly, depend on the machinery of primary emotions, since natural selection rarely evolves entirely new mechanisms from scratch, but tends to conserve devices that have worked so far. Now, in view of the fact that secondary emotions involve connections that are not inherited but derived from experience, the neural network required to support them would have to be extended beyond structures in the limbic system to include the prefrontal and somatosensory cortices.

Somatic markers, which can now be more accurately defined as 'special instance[s] of feelings[6] generated from secondary emotions' (Damasio 1994, p.173), also depend for their acquisition on the prefrontal cortices. Prefrontal cortices – Damasio believes – are involved in categorizing things and events according to personal relevance, in the sense that 'prefrontal networks establish dispositional representations for certain combinations of things and events' on the basis of each individual's life experience and as determined by their survival value.

What makes the prefrontal sectors prime candidates as neural substrates for somatic markers lies in the fact that they are convergence zones for signals about all the relevant information, namely:

(a) signals from sensory regions about the external world (which includes social conventions and ethical rules);
(b) signals about innate biological preferences from the internal preference system. Whose neural basis mostly consists of 'innate regulatory dispositions,[7] posed to ensure survival of the organism';

6 This is how Damasio (1994, p.145) defines feelings: 'If an emotion is a collection of changes in body states connected to a particular mental image that have activated a specific brain system, the essence of feeling an emotion is the experience of such changes in juxtaposition to the mental images that initiated that cycle. In other words, a feeling depends on the juxtaposition of an image of the body proper to an image of something else, such as the visual image of a face or the auditory image of a melody.'

7 We are here talking of bio-regulatory sectors of the human brain, including the neuro-transmitter nuclei in the brain stem, and the basal forebrain, as well as the

(c) signals from the somatosensory cortices about the continuous succession of body states brought about by knowledge relative to the external world and innate preferences.

In addition to receiving all the above signals, the prefrontal cortices are also 'directly connected to every avenue of motor and chemical response available to the brain' (Damasio, 1994 p.183).

Somatic markers depend for their functioning on two alternative mechanisms, one involving the body, the other in which the body is bypassed. The first, or basic, mechanism has the body assume a particular state under the influence of the prefrontal cortices and amygdala. This state is then signaled to the somatosensory cortex. In the alternative[8] or 'as if' mechanism, 'the prefrontal cortices and amygdala merely tell the somatosensory cortex to organize itself in the explicit activity pattern that would have been assumed had the body been placed in the desired state and signaled upward accordingly' (Damasio 1994, p.184).

Both mechanisms can operate overtly or covertly: irrespective of whether the somatic state or its surrogate is the focus of attention, the underlying mechanism is activated. Thus, if we are dealing with the basic mechanism operating in a covert manner to signal a negative outcome, the corresponding explicit imagery would be generated, but the ensuing body change would not be perceptible. This, however, would nevertheless result in an inhibition[9] of the tendency to act, or the actual enhancement of the tendency to withdraw, by virtue of an attentional mechanism.

Whether they operate overtly or covertly, somatic markers cannot carry out their biasing job unless they can draw upon the mechanisms of basic attention and basic working memory, both of which are necessary to the display of the categorized knowledge we

amygdala, the anterior cingulate, and the hypothalamus (cf. Damasio 1994, p.181).

8 Having this alternative mechanism would explain that so many decisions can be taken on a day-to-day basis, that are not accompanied by feelings.

9 This inhibition is effected via the regulatory neural circuits located in the brain core (Damasio 1994, p.187).

possess concerning situations, options and outcomes (Damasio 1994, pp.196–8). The former serves to maintain a given mental image in consciousness, presumably by enhancing the corresponding neural activity pattern; the latter, to hold separate images for a period of hundreds to thousands of milliseconds, during which the supporting representations are reiterated. Basic attention and working memory would also be required during the reasoning process following the operation of somatic markers. Damasio's proposal is that:

> a somatic state, negative or positive, caused by the appearance of a given representation, operates not only as a marker for the value of what is represented, but also as a booster for continued working memory and attention.

One last aspect of the above decision-making process has to do with 'the critical problem of creating order out of parallel spatial displays', in response to the normal requirement of cognition and movement.[10] Damasio subsumes the need for decision-making under that of creating sequential order out of a concurrent display of possibilities, in the event, of alternative options available from the store of factual knowledge. In such a scenario, somatic markers acting as ranking criteria (based, at any one time, on innate and acquired preferences) would assign different degrees of 'attentional enhancements' to different contents.

Before we return to the knowledge source problem, let us, for further reference, summarize the main features of Damasio's model and consider what further level of automatization could, in principle, be carried out. Damasio's model includes:

10 'The brain's design only permits, at any one time, a limited amount of conscious mental output and movement output. The images which constitute our thoughts must be structured in "phrases," which in turn must be "sententially" ordered in time, just as the frames of movement which constitute our external responses must be "phrased" in a particular way and those phrases placed in a particular "sentential" order for a motion to have its desired effect' (Damasio 1994, pp.198–9).

- A vast repository of factual knowledge consisting, among other things, of categories of situations paired with categories of response options, in turn paired with attendant categories of outcomes. (As a broad-based knowledge, it also includes 'facts about objects, persons and situations in the external world [and] facts and mechanisms concerning the regulation of the organism as a whole.' (Damasio 1994, p.83–4))
- A biasing system, which allows response options to be pre-selected, based on the operation of somatic markers, or somatic states with which individual experience has tagged positive or negative outcomes. Whether or not those somatic states are perceptible to the organism, outcomes thus tagged will accordingly affect the tendency to act or to withdraw.
- A reasoning process, which takes as input the shortlist of options provided by somatic markers.

Both somatic marker device and reasoning process require the supporting mechanisms of basic attention and basic memory to maintain given mental images in consciousness for an extended period of time for comparison purposes.

As it stands, Damasio's model already presents a clear departure from the basic model in terms of automatization. Thus, somatic markers provide an automatic pre-selecting device that considerably simplifies the work of the reasoning process. Furthermore, as a result of development, the device can bypass actual somatic states and rely on 'symbols' of somatic states or 'as if symbols' (Damasio 1994, p.184). A yet further level of automatization is exemplified by the fact that neither the basic nor the alternative mechanism need be the focus of attention.

What further plausible levels of automatization could be envisaged, which would be susceptible of implementation in the language version of decision making? The one obvious area still open to automatization is the reasoning process whereby the final decision is made. Increasing the level of automatization of this process would entail constraining the way it operates. Indications that this possibility may be realized lie in the existence of argumentative discourse markers,[11] whose meaning is construable in terms of encoded reasoning patterns (which would implement a top-down form of interpretation at the text level, providing specific frames of interaction and relationships requiring instantiation).

An alternative to automatizing the reasoning process itself consists in reducing the number of options to be compared, which appears to be feasible, if lexical items of the type under consideration are something to go by; as will be recalled, 'rich', for example, appears to be encoded with a weak constraint, whereby its occurrence evokes scenarios featuring practices normally associated with rich people. A similar function is performed by lexical items[12] such as *reckless* vs. *brave*, or *cowardly* vs. *prudent* that include, as part of their core meaning, an implicit negative or positive judgment or attitudinal component, accompanied by a stronger level of constraint on continuation (for further details, cf. Bruxelles, Ducrot and Raccah 1993). This latter type of encoding comes close to somatic marking, as the use of these lexical items can negatively or positively predispose the interpreter towards the object being referred to. Note that the very fact of representing an option (e.g. by providing the conclusion to be reached) constitutes a form of imposed selection to the exclusion of all others (H might have in mind).

Extending Damasio's repository of categorized factual knowledge

Why this store of knowledge should be extended is because the situation-response pairs corresponding to argumentational meaning, though clearly unrelated to the regulation of the organism as a whole, do not appear to be based on individual experience either. Thus, while individual experience may cause some of us strongly to link stinginess with wealth, and think twice before approaching wealthy people for

11 See Nyan 2003, 'From procedural meaning to processing requirement', in *Discourse particles*, Kerstin Fisher (ed.), to appear.

12 Unlike what happens with complex categories, where the tagging concerns response-options, in the cases at hand, it is the initial situation (or an element of it) that appears to have been tagged.

financial help, the connection between wealth and buying power is something that directly stems from what we understand wealth to be about, not merely conceptually, but also in terms of the kind of practice wealthy people are able to engage in. The same point can be made with regard to beauty and attractiveness, and what problems entail in terms of the level of effort required to deal with them.

Concerning those pairings, which appear to underlie preferred continuations, the following questions arise: what type of knowledge they are based on? How do any given two representations come to be paired? In the cases discussed by Damasio, we are dealing with factual knowledge in the form of categories of contingently related situations, which become linked through individual experience. By contrast, the categories of situations exemplified in intrinsic topoï (before they become intrinsic through encoding) stand in a relation that appears to arise from forms of know-how ranging from shared cultural and social practices, to something more fundamental, namely a set of abilities we share as human beings inhabiting the same physical world. These various forms of know-how pertain to what Searle (1983 and 1995) calls 'the Background' (which consists of different layers, with a local or cultural Background at one end, and a deep Background at the other).

Searle's theory of the background

Searle (1983, p.143) defines the Background as 'a set of non-representational mental capacities that enables all representing to take place.' The forms of representing being referred to include meaning, perception, action, mental states with a direction of fit, such as belief, desire, intention. What is meant by 'capacities' are certain kinds of know-how about 'how things are and how to do things', which include 'various skills, abilities, preintentional assumptions and presuppositions, stances and nonrepresentational attitudes'; to which should be added 'dispositions, tendencies, and causal [neurophysiological] structures generally' (Searle 1995, pp.129–30). Those capacities are nonrepresentational in the sense that knowing how to activate them in performing skills does not involve having them as objects of representation (or intentional objects) of a mental state such as belief (see also note 3, Chapter 3). For an example of how a preintentional Background provides the necessary conditions for the functioning of intentional contents, consider what must be brought to bear on our understanding of the processes described by the literal meaning of 'cut' in 'Sally cut the cake' and 'Bill cut the grass', which differ both in terms of what the agents do and what they use to do the cutting. The lack of such a background with relation to 'Liam cut the trail' makes 'cut' somewhat difficult to grasp despite the fact that it is also used literally.

Between forms of representation (such as meaning, perception, intentional states, action, including performing and interpreting speech acts), which are intentional and the Background, which is preintentional, Searle postulates a network of intentional states, the Network. This network of intentional states is what enables:

> each intentional state [with a direction of fit] to determine its conditions of satisfaction. Thus. one cannot form the intention to run for the Presidency of the United States without having certain beliefs (e.g. the United States is a republic, [...]. That the candidate of the two parties vie for the nomination of his party, [...], that voters vote for him.

The above intention 'can only have the conditions of satisfaction that it does, and thus can only be the intention it is because it is located in a Network of [other] beliefs and desires', which also include subsidiary intentions as well as other psychological states. Spelling out what the Network consists of would be an impossible task in view of the fact that 'much, perhaps most the Network is submerged in the unconscious'; furthermore, if one were to attempt to carry out the task, one would 'eventually reach a bedrock of mental capacities that do not themselves consist in Intentional states (representations), but nonetheless form the preconditions of Intentional states' (Searle 1983, p.141).

Thus, the man in the previous example cannot form the intention he does, unless, in addition to having a whole network of intentional states, he also, for example, takes for granted that 'elections are held at or near the surface of the earth', in the sense that he would be very surprised if that were not the case, rather than in the sense of having such a belief. This bedrock of mental capacities into which the Network shades is the Background.

From background knowledge to complex categories

As will be recalled, the complex categories discussed by Damasio (CCs1) arise from the pairing of types of situation calling for a decision (e.g. those featuring falling objects), with types of response option (e.g. ducking or catching), which have been marked as advantageous or dangerous on the basis of individual experience. This linkage between categories of situations and categories of response-options forms the basis for a know-how that facilitates decision-making, namely the ability to access the desired response options automatically, given a certain type of situation.

By contrast, the complex categories under consideration (CCs2) arise from the fact that entities (or events) are inextricably associated with practices and abilities that, as members of the same species and societies, we have developed in dealing with the types of entity they are. While this association provides us with the appropriate skills to engage with them, it is also part and parcel of a deep understanding we have of the kind of entities they are. In terms of their relation to the decision-making situation, we have an even more marked contrast. In the case of CCs1 the situation calling for a decision is subsumable under the category of situations corresponding to the first member of the complex category, and the type of response it requires is available from the second member of the complex category, or the set of options that instantiates it. With CCs2, the situation calling for a decision is communicative and consists of the fact that a representation of a situation (corresponding to the first member of the complex category) has been made with a communicative intent; as for the response option, it is also a linguistic event, that of following up the first representing event with a continuation expressing an attendant situation, or identifying such a continuation. An example of the first kind of decision-making situation is one where the (real life) situation calling for a decision (e.g. having a bully among one's colleagues) is recognizable as a type of situations one has successfully handled in the past in a certain way (e.g. by refusing to be intimidated). In such as situation the corresponding categorized knowledge will provide a quick response option. For an illustration of the second kind of decision-making situation, consider:

(7) A: How come we haven't seen you around lately?
B: This whole new development at work has been a real pain, so well, you know…

In this case, the situation calling for a decision is the representation of the situation at work; which representation is made with the communicative intent of getting A to understand what is meant by 'so well, you know'. A's background knowledge may be what enables her immediately to grasp what having to deal with a situation which is a pain entails, however, the reason that A knows what is expected of her lies in her linguistic skills, in virtue of which she recognizes what type of decision the communicative situation calls for, has access to a set of response options, and, in this particular case, can immediately reach a decision, thanks to a constraint that constitutes an encoded link to a preferred option. Such a constraint would be based on a complex category derived from A's nonlinguistic background knowledge.

But how does background knowledge fit in with Damasio's contingently acquired factual knowledge? The very first thing that needs to be said is that we are not dealing with a set of know-how as opposed to a know-that. As will be recalled, the factual knowledge Damasio is concerned with is categorized, and the link between types of situation and types of response-options, which exhibits varying degrees of automatization, provides the basis for a certain know-how people rely on in dealing with situations requiring a decision. So the relevant difference is only one between two types of know-how, as based on their mode of acquisition. The know-how that interests Damasio is acquired on the basis of personal experience, while background knowledge stems from collective experience, in the sense of the same type of experience occurring to individuals sharing the same social and physical environments, as well as the same biological make-up. In other words, individual skills susceptible to modification through exposure to further experience would be found on the same continuum as a better entrenched cultural background into which they also shade; and this cultural background in turn also shades into the deep background we all share.

However, viewing things in terms of a continuum is misleading, unless one bears in mind what Searle says in respect of the relationship between the Background and the Network, namely that the former permeates the latter. Consider again the situation involving the falling object. Irrespective of what our personal experience has conditioned us to regard as an advantageous response, we cannot even begin to categorize the situation as potentially dangerous or envisage an appropriate range of responses, unless we possess the relevant background assumptions about what effects a gravitational field has on the physical objects situated within it.

But, do we have a similar relationship between cultural practices and individual know-how? Insofar as the kind of know-how involved in cultural practices (i.e. how to handle given situations in a way that is sanctioned by the culture one lives in) does not provide the enabling skills required in the exercise of individual know-how, the answer has to be in the negative. Nonetheless, it would not be appropriate in this case either simply to think in terms of a continuum. Indeed, what makes the choice of a given response a matter of cultural, rather than personal preferences, is that society's approval or disapproval has had a determining role in its acquisition: as part of our knowledge of the external world, awareness of social and ethical rules would trigger somatic states, which mark the response in question as acceptable or not. However, this is not to say that personal preferences (masochistic or otherwise) cannot coincide with a sense of social obligation, or that the same response cannot be accompanied by conflicting somatic states brought about by conflicting experiences (e.g. feeling both elation and dread when engaged in breaking rules).

So the way deep background knowledge fits in with the set of individual skills that arise from personal experience is by providing those enabling capacities that make the activation of these skills possible, while the cultural background may be seen as a parallel set of skills to individual know-how in situations subject to social and cultural norms.

The fact that background knowledge underpins individual know-how does not quite solve the problem we have here however. What we said earlier about communicative skills entails that the use that is made of background knowledge in the exercise of those skills is not one involving the activation of a know-how, but rather the representation of a know-that. Consider (7) again. The decision required of A is not that she uses her background knowledge to sort out the crisis at B's work place, but rather that she derives from it the correct scenario that will enable her to understand what B means (and in what way his response is relevant to her question). Now, the reason why we have this problem is because, if – as Searle claims – background knowledge is nonrepresentional, it is unclear how it could have made its way into complex categories, which are representational? As we saw in the previous chapter, although a know-how can only be activated (in the performance of a skill), not represented, it appears to be possible to have an equivalent know-that, for instance in the form of scenarios featuring people exercising the skills in question. If we choose to go down this route, can we still talk about complex categories arising from background knowledge (as we obviously want to do)? The answer to this question would appear to lie in the role of dispositional representations. Dispositional representations, to reiterate Damasio's definition (1994, p.102) 'exist as potential patterns of neuron activity in small ensembles of neurons [or "convergence zones"]; that is, they consist of a set of neuron firing dispositions within the ensemble'. As such they constitute our full repository of knowledge, both innate and acquired. Those on which innate knowledge is based are located in the hypothalamus, brain stem and limbic system and control processes required for survival. The ones that concern us, and on which acquired knowledge is based, are located throughout the higher-order association cortices (in occipital, temporal and frontal regions). Among the latter set of dispositional representations, some 'contain records for the imageable knowledge that we can recall and which is used for movement, reason, planning, creativity; and some contain records of rules and strategies with which we operate on those images' (Damasio 1994, p.105).

Let us now return to the question of how there can be a know-that corresponding to background knowledge. Assuming that background knowledge is acquired in the same way as individual know-how, that is, through experience, then the perceptual images corresponding to that experience may still be susceptible to re-activation in the early sensory cortices through the firing of relevant dispositional representations. Now, what if these images have become too faint to be re-activated? If we are talking about cultural know-how as opposed to deep background knowledge, these perceptual images may be primed by or consist of scenarios involving other people, which would have acted as motivators in the acquisition of the know-how. Should there be no early perceptual images that can be re-activated, a possible – albeit less plausible – alternative could be provided by a sequentialization process: background knowledge, construed as a case of 'parallel spatial displays', would somehow become sequentialized in response to the normal requirement of communication, which has to take place in time, and is presumably subject to a constraint mentioned earlier: as will be recalled, the brain's design is such that it 'only permits, at any one time, a limited amount of conscious mental output and movement output' (Damasio 1994, p.190). This alternative seems less plausible, unless there is a process whereby a know-how can be 'converted' into a know-that, as opposed to having an equivalent know-that which is independently acquired.

The upshot of the above discussion is that not only does the individual know-how arising from Damasio's categorized factual knowledge appear to exist on the same continuum as Searle's background knowledge, but the complex categories it is made up of shade into the complex categories that correspond with background knowledge. The linguistic skills we are concerned with, which are subsumable under decision-making skills, would draw upon the latter type of complex category to reach a decision about the upcoming text.

The interpretative problem: argumentation, categorization and divergent thinking

In the previous section I hypothesized that the type of communicative situation that gives rise to constraints on continuation, constitutes a case of decision making, as based on the model advocated by Damasio. This linguistic version of decision making relies on a source of knowledge which is an extension of Damasio's repertory of categorized factual knowledge, and exhibits a further level of automatization relative to this model, inasmuch as constraints on continuation largely obviate the need for an elaborate selecting process.

In this section we turn our attention to what may underlie the encoding of such constraints. As announced earlier, our primary concern is the type of interpretative process that could have facilitated this encoding. We will be arguing that divergent thinking has a crucial role in creating this interpretative shortcut, which in turn paved the way to the encoding.

A word about the organization: our specification of this interpretative process begins with a description of the type of communicative act it serves to identify, namely the act of argumentation. Indeed, the latter arguably involves categorization and reference to complex categories – an essential feature, if the corresponding interpretative process is to lead to the encoding of constraints on continuation.

Acts of argumentation

Argumentational data, as will be recalled, involve constraints on continuation that appear to stem from the conceptual structure of certain lexical items (among other things). Following AWL, we will assume that this conceptual structure came to incorporate such constraints via processes akin to delocutivity, whereby repeated occurrences of these lexical items in certain contexts or co-texts would, over an extended period of time, result in linguistically encoded links.

(It goes without saying that this kind of conceptual change could also be brought about by experiences other than linguistic.) Among possible linguistic uses that could have triggered this encoding process, one in particular recommends itself for consideration, namely the performance of acts of argumentation,[13] as based on premise/conclusion pairs such as (8)–(9):

(8) Ludwig feels neglected (P).
(9) He requires your attention (Q).

Inasmuch as the aim of this chapter is to argue that argumentational meaning results from a further application of divergent thinking, it is important that we should be able to show that:

(a) acts of argumentation are distinct from illocutionary acts, as categorized by Searle (1979);
(b) they present properties of their own, which would support the thesis of a different application of divergent thinking.

Consider again the premise/conclusion pair in (8)–(9), where P stands for the propositional meaning in (8), and Q for that in (9). It could be argued that viewing (8)–(9) as allowing an act of argumentation – rather than an illocutionary act followed by a communicative act – is a matter of theoretical perspective (i.e. AWL vs. Speech Act Theory); that adducing a premise in support of a conclusion amounts to performing a complex speech act construable in terms of:

13 There is a clear heuristic value in viewing the argumentational phenomenon in relation to acts of argumentation. Acts of argumentation can be correlated with neurobiologically rooted models of decision making, as well as with other types of meaning arising at the text level, with which they share similar processing problems, with what that entails in terms of strategies available for co-option. Other instances of language use that may be conducive to the emergence of argumentational meaning do not offer the same possibility of correlation.

(i) a commitment on S's part to the truth of the relevant proposition (P), or to a words-to-world direction of fit between P and the situation of reference;
(ii) a link between S's commitment to P and the fact that H is being placed under an obligation to do something, in the event, to accept the conclusion (Q) expressed in (9).

While, roughly speaking, correct, this analysis fails to take into account an important aspect, which has to do with the nature of the obligation placed on H: Viewing (8)–(9) as a communicative act (in the sense of Searle 1979) – one based on a prior assertion – entails that this obligation derives its legitimacy from S's commitment to the truth of P. This, however, makes no provision for the relation between P and Q: while H can conceivably accept that P is true on the basis of S's commitment to P's words-to-world direction of fit, this does not quite warrant his acceptance of Q, as a conclusion to be drawn from P. To be able to accept Q as such a conclusion, it is necessary for H to assume that Q is subsumable under a set of conclusions that can be drawn from a set of premises under which P can be subsumed. In other words, the main focus of an analysis in strict speech act terms is the interactional level, which leaves the relationship between propositions insufficiently accounted for.

To cater for this relationship one can appeal to Anscombre and Ducrot's notion of topos (or shared reasoning principle). A reanalysis of the above example in terms of topoï runs as follows: how S purports to get H to accept her conclusion on the basis of the argument put forward is by referring to a shared topos (T) to the effect that, *[the more someone feels neglected, the more attention he should receive]*. The obligation now being imposed on H is one of consistency: if T is a shared topos, then H is consistency bound to accept that Q should be drawn from P.

Though a clear improvement on the previous solution, the proposal under consideration fails to explain how S's commitment to P's truth fits into the picture. That this commitment has a role to play in the performance of an argumentative act is clear from the fact that the latter can be invalidated by denying that P is the case, even though accepting the truth of P does not quite get one to the argumentative stage, as attested by the appropriateness of (10) in response to (8)–(9):

(10) Ludwig feels neglected but I have other priorities.

The appropriate type of solution should, as a matter of course, preserves the idea of S's commitment to P's truth; however, it should also – more importantly – explain how this commitment interfaces with an act of argumentation on the higher level, in the sense of how S's commitment to P being the case relates to her referring to a topos, as well as to how this commitment gives rise to an obligation for H to accept the conclusion. To get a clearer picture of the difficulty involved, one can describe the gap separating the two levels in terms of direction of fit. Thus, on the one hand, S's commitment to P being the case corresponds to a words-to-world direction of fit; on the other, S's reference to a topos corresponds to a world-to-mind direction of fit. Thus set out, the two directions of fit show the gap to be a matter of 'words' being left unaccounted for in the process linking the lower to the higher level of analysis.

My proposal is that acts of argumentation should be reanalyzed as involving a categorization process: unless the situation of reference falls under the category of situations featured in the antecedent of the topos, this topos – a type of complex category – will not be applicable in the context of utterance. On this analysis, acts of argumentation take as their starting point a different aspect of P, namely P in a categorizing capacity, as determined by its relation to a topos; previously this starting point was provided by P in a representational capacity, as defined by its relation to the situation of reference. Thus, in saying (8), with an argumentative intent, S is, as a matter of course, presupposing two things:

(a) that P is a true representation of a state of affairs in the world;
(b) a shared topos T, *[the more someone feels neglected, the more attention he should receive]*, which, metalinguistically, may be glossed as *[the more a given situation is a member of a set of situations {sit1}, the more it is conducive to a situation belonging to another set of situations {sit2 }]*.

In addition to presupposing the above, S – more importantly – is committing herself to P being an appropriate category judgment of the situation of reference. P, in this case, is envisaged under its categorizing aspect, and the features it picks out of the situation of reference are those in virtue of which this situation would qualify as a member of {sit1}. On this analysis, there is no separate act of reference to the topos: viewing P in a categorizing capacity necessarily involves referring to some complex category; furthermore, S's commitment to P as an appropriate categorization of the situation of reference necessarily presupposes a topos that defines what constitutes an appropriate categorization of that situation.

Turning now to H's obligation, S's commitment to P's words (-to-world) -to-mind direction of fit is what underpins the obligation she places on H. Thus, in saying (8) in the context at hand, S is committing herself to P being the correct category judgment for the situation of reference, and in so doing, places H under an (illocutionary) obligation to accept P as such, or, alternatively, to accept the situation of reference (as featured in P) as a member of a category of situations {sit1}.

In view of the fact that reference is being made to a shared topos – whereby {sit1} is linked to a further category of situations {sit2} – H's previous obligation doubles as an obligation of consistency: assuming that his mental context includes T as the prevailing topos, in accepting the proposed categorization of the situation of reference, H is consistency bound to regard this situation as conducive to a state of affairs presenting the membership criteria of {sit2}.

The fact that, over and above her argumentative intent, S could also be attributed an illocutionary intent – that of getting H to bring about the state of affairs described in Q – does not entail that acts of argumentation are quite subsumable under one of the known classes of illocutionary acts. After all they do exhibit three features of their own:

(1) P, the propositional meaning on which the premise is based, is to be viewed under a categorizing aspect, as opposed to a representational one (as would be the case with assertive acts). In a categorizing capacity, P does not merely purport to represent the

situation of reference, it purports to do so under those features that make it possible for its membership status to {sit1} to be assessed. In other words, the relevant direction of fit is one involving words (or rather words-to-world) and mind, as opposed to merely words and world. As for the purpose of the act, it is twofold:

(a) to commit S to a words (to-world)-to-mind direction of fit;
(b) in so doing to place H under an obligation to accept that P, as a category judgment of a state of affairs in the world, conforms to mind (i.e. to a category of situations {sit1}). (On a strict illocutionary analysis, the purpose of the act is to commit S to P being the case, that is, to a words-to-world direction of fit (as in assertive acts), thereby placing H under an obligation to accept the same direction of fit.)

(2) The nature of H's obligation: while S's commitment to P's words (-to-world) -to-mind direction of fit may serve as a validation for H's (illocutionary) obligation to accept the same direction of fit for P, H's obligation to accept the conclusion in Q is a matter of consistency: it arises from his own commitment to P as an appropriate category judgment for the situation of reference, and from his adherence to an alleged shared social construct, or topos. On a strict illocutionary analysis there is no provision for why H's acceptance of P (or of its words-to-world direction of fit) should lead to his acceptance of the conclusion. In terms of domains of reference, if in both cases the existence of obligations point to an intersubjective domain arising from the domain of discourse, reference to shared social constructs introduces an intersubjective domain of a more global kind, one comprising social constructs in the form of complex categories (this domain corresponds to Searle's local Background).
(3) The final purpose of the (overall) act – as based on the expression of both premise and conclusion – is, unambiguously, H's acceptance of the conclusion. In other words, the final (expressed) purpose of the act is H's acceptance of Q as a conclusion, or as having a words-to-mind direction of fit. The

fact that the context also licenses the interpretation that Q expresses the conditions of satisfaction of a future state of affairs H is expected to bring about does not entail that the corresponding directive falls within the purview of the act. Such a directive would be based on the assumption that S possesses an actual intention to get H to bring about a certain state of affairs, which is distinct from his expressed argumentative intention. This distinction in purpose between the two types of act also entails a distinction in terms of the target domain of reference (or domain concerned with the expected change). If getting H to accept a given conclusion amounts to securing his commitment to a certain proposition, then the appropriate domain is a subjective domain, namely that of his mental states. By contrast, this domain would appear to be the external world in the case of a directive, in view of the nature of the corresponding goal. An illocutionary analysis, though capable of distinguishing the two types of purpose, only acknowledges one type of act (Searle 1979, p.13). Thus a directive will place H under an obligation to do one of two things: to bring about a state of affairs in the external world, or to accept a conclusion, which is a matter of bringing about a mental state. From the fact that no status as such is accorded to acts of argumentation nothing need follow in terms of distinctions to be made with respect to shifts in domains of reference. Now, as it happens, such shifts – as will be seen further on – are indicative of attentional shifts, with what the latter entail in respect of changes of levels of analysis, perspective and scope; and attentional shifts are among the prerequisites to the emergence of new meanings at the text level.

Granted that there are such things as acts of argumentation – differing from bona fide illocutionary acts with regard to the aspect under which propositional meaning is envisaged, direction of fit, S's commitment, H's obligation, domain of reference, and constraints on continuation – they would have given rise to such constraints through the persistent linkage of the same pairs of contents acting as premises and conclusions. Among the various functions continuations can assume – especially if they are predictable – an obvious possibility is for them to express conclusions. What constitutes a conclusion from an inferential perspective is a continuation from the perspective of the co-text, one predictable on the basis of a successful act of argumentation.

Acts of argumentation and decision making

Acts of argumentation, as just described, may be seen as attempts to induce in H a decision-making process, whereby the correct interpretation of S's intention is reached.

In light of what was said previously, two levels of decision making need to be distinguished, one linguistic, the other nonlinguistic. From a linguistic or communicative standpoint, the situation calling for a decision is the utterance or sequence of utterances produced by S with an argumentative intent, and the decision this situation calls for is a linguistic response of an appropriate kind (i.e. one of identifying and accepting a certain conclusion). H knows what kind of response is appropriate because he possesses a certain know-how about communicative situations, in the form of complex categories consisting of categories of communicative situations linked to categories of appropriate responses. These complex categories would have come about in the same way as Damasio's complex categories, that is, through exposure to experience: appropriate behaviour would have been socially re-inforced, inappropriate behaviour discouraged, with each type producing relevant somatic states. Whether or not, as adults, we still feel a twinge of guilt (assuming that we ever did) when we fail to respond to speech acts in the expected manner, the knowledge of such a link is there, which, perhaps not surprisingly, is also that of an obligation imposed by speech acts.

Now, knowing what kind of response is expected does not quite tell one what that response should be. Knowing that one is to identify and/or endorse a conclusion merely tells one about the nature of the appropriate stance is, not what the intentional object of that stance, or the content of the conclusion is. This is where the second level of decision making comes in: in order to identify the content of the conclusion, H has to access relevant complex categories about nonlinguistic situations such as the one described in P.

But before proceeding to our next point, we must clarify in what sense we are dealing with a case of induced decision making. The reasons for it should be obvious:

(1) The situation calling for a decision is one of S's making.
(2) Where a whole range of options would normally be made available for consideration, we are now left with just one, as linguistic representation only allows one option to be represented at a time.
(3) The process whereby reference to specific types of complex category is made can be automatized to varying degrees, if the lexical items involved happen to be encoded with the relevant constraints.

However, for this encoding to take place, the interpretative process needs, as a matter of routine, to be able to:

(a) switch rapidly from one aspect of P to another (i.e. from P in a representational capacity, to P in a categorizing capacity);
(b) to have immediate access to the relevant store of complex categories.

On our hypothesis divergent thinking is the underlying capacity that biases the interpretative process to fall into the above routine, thereby creating an interpretative shortcut that allows constraints on continuation to be encoded.

Argumentational meaning and divergent thinking

On the basis of the previous analysis we are now in a position to specify on what grounds argumentational meaning, as arising from acts of argumentation, can be put down to a further application of divergent thinking. But first, let us reiterate what criteria we take to be indicative of a susceptibility to divergent thinking in general, and how these criteria can be further specified in terms of the domain of application.

As will be remembered from Chapter 1, inasmuch as divergent thinking involves an attentional shift from lower to higher level, a general requirement is that possible candidates for lower and higher levels should belong to a coherent conceptual domain[14] and the relevant objects of perception within it should be susceptible to a construal in terms of such levels. As for the criteria themselves, four were proposed:

(1) These levels should stand in a hierarchical relationship, as defined by the domain of application. As will be recalled, divergent thinking typically underlies tasks, such as symbol learning, which involve subordinating elements to one another.
(2) The lower level should be seen under a different aspect to the original, when viewed in terms of the higher level. This stipulation reflects the change of perspective entailed by the shift upwards, whose purpose is to sample allocentric, rather than egocentric, alternatives.
(3) The directionality of the shift should be from a domain of reference that is less abstract to one that is more abstract.[15]
(4) There should be an identifiable problem for which shifting attention to the higher level provides a better solution.

14 This conceptual domain, as will be remembered, need not correspond to a dimension of experience, as normally construed, but could be theory-bound.

15 Regarding what constitutes an abstract domain, Croft (2000, p.260), following Langacker 1987, has this to say: 'abstract domains are [themselves] components of a very large number of concrete domains. For example, almost all domains make some reference to time or scales in defining the concepts profiled in them (any domain involving events and gradable properties respectively), and an enormous number involves some reference to spatial concepts in their structure (any domain involving physical objects or motion). It is more precise to think of domains as being themselves multidimensional [...] the "dimensions" that are found in the largest number of domains are the most abstract ones.'

As previously, the above criteria are susceptible of further specifications in terms of the domain of application under consideration, namely that of meaning at the text level, with propositional meaning and speech acts providing possible reference points for lower and higher levels

Before we turn to acts of argumentation, a quick reminder of how illocutionary acts meet these criteria would be useful:

- The hierarchical relation between lower and higher-level elements is determined by the logical structure of intentionality (cf. Searle 1973, pp.4–26), in that the propositional content one has at the lower level 'becomes', at the higher level, the intentional content of a mental state, whose expression is the corresponding illocutionary act.
- The change in aspect required by the second criterion is inherent in what has just been said: the same propositional content that is envisaged in a representational capacity at the lower level is seen, from the higher level perspective, to be the intentional content of a mental state and of the speech act that expresses that mental state.
- The third criterion – about the directionality of the shift – is satisfied by the fact that, parallel to the functional shift undergone by the propositional content, there is a shift in domains of reference, with the source and the target domains being instantiated by the external world and a subjective world (including mental states), respectively.
- The fourth criterion is also met: there is an interpretative problem for which divergent thinking provides a better solution, namely the identification of the object of communication, in the event, the intentional state, of which the illocutionary act is the expression.

Acts of argumentation and divergent thinking

Acts of argumentation, which we take to be likely – albeit by no means exclusive – sources for argumentational meaning, also meet the above criteria, but do so in a way that sets them apart in terms of application of divergent thinking.

Concerning the first criterion, insofar as acts of argumentation are amenable to an analysis in terms of the structure of intentionality, it would appear that we have the same two elements standing in a hierarchical relation, namely, a propositional meaning, at the lower level, and the expression of a mental state at the higher, with the propositional meaning providing the mental state with an intentional content.

While this might do as a first approximation, it does not provide enough distinctions to separate acts of argumentation from bona fide illocutionary acts, as arising from a different application of divergent thinking. Consider again:

(8) Ludwig feels neglected.

Viewing (8) as marking an act of argumentation, as opposed to an assertion, involves a different construal of the intentional content. Where an assertion would have, as sincerity condition, the belief that the propositional content (as a purported representation of the world) conforms to a certain state of affairs, in the case of an act of argumentation, the intentional content is two-tiered, so to speak: the bottom tier corresponds to the belief that P is a true representation of a state of affairs in the world, the top tier, the belief that P, as a true representation of that state of affairs, presents the latter as possessing certain features, those relevant to the assessment of its membership status with regard to a certain category of situations. The top tier is the one S's commitment is concerned with: in saying (8) with an argumentative intent, S commits herself to P being an appropriate categorization for a certain state of affairs; her commitment to P as a true representation is presupposed.

The second criterion, as will be recalled, is about the aspect under which the lower level interfaces with the higher level. Why there should be a change of aspect follows from the fact that the upward attentional shift entailed by divergent thinking is concerned with allocentric, rather than egocentric solutions. In other words, not only is propositional meaning integrated to a higher unit of meaning, thereby becoming the intentional content of a mental state, but the relevant aspect of P must be different from the one that matters at the lower level.

Before proceeding further, let us remind ourselves in what way a change of aspect correlates with divergent thinking. At the text level, the discrepancy between the means of expression (i.e. linguistic structures) and the intended meanings becomes more pronounced. As a consequence, processing efficiency is at risk of being seriously impaired unless a shortcut can be found to provide a direct link between the strings under consideration and the complex meanings intended by S. (That the solution should take the form of a shortcut of this kind is consistent with what is known of the tendency towards automatization characteristic of processing systems). Now what stands in the way of this discovery is that these strings already have a meaning, propositional meaning, in virtue of linguistic conventions. In other words, the linkage cannot occur unless this meaning is set aside or, rather, the attendant analysis inhibited, to allow for an attentional shift to the higher level. This, we recall, can be achieved through divergent thinking. So, given that this shift also constitutes a shift from egocentric to allocentric information – construable, in this case, as information concerning the external world and information concerning S's mental state, respectively – where does the change in aspect come in? The answer lies in the intentional structure of meaning and the logistics involved in finding a quick access to higher-level meanings on the basis of strings conventionally associated with lower level meanings. As we know, propositional meaning provides the mental state expressed by higher level meaning, with its intent-ional content, so the ability to gain quick access to this higher level meaning on the basis of the lower level meaning is dependent on the ability to view P under the appropriate aspect. Consider (11), to which is associated a propositional meaning P, expressing a certain state of affairs:

(11) They went off to Half-Moon Bay.

(11)'s susceptibility to being interpreted as marking a speech act would be dependent on an ability on H's part to view P in terms of its role at the higher level, which is to provide the intentional content of a belief. This ability thus involves switching from one aspect of P to another, that is, from P expressing a state of affairs to P expressing the content of a belief that the state of affairs is the case. This switch is what underlie the shortcut.

Returning now to acts of argumentation, what aspects of P are involved in reaching an argumentative interpretation of a string such as (8), on the basis of the lower level structure? In order to interpret (8) (repeated below):

(8) Ludwig feels neglected.

as marking an act of argumentation in support of a conclusion C (to the effect that H should do something about it) one must be able to set aside P, as propositional meaning – representing a state of affairs – and view it as providing with an intentional content, not only the belief that P is the case, but also this further belief that P being the case is a sufficient basis for C (In more technical terms, this further belief is that P provides an appropriate category judgment of the situation of reference, whereby the latter is shown to have satisfied the membership criteria for a set of situations {sit1}; {sit1}, is assumed to be linked to another set of situations {sit2}).

Reaching an argumentative interpretation of (8) would thus require two switches:

- One from P in a representational capacity (i.e. as expressing the fact that Ludwig feels neglected) to P in a categorizing capacity, or Pc. (In virtue of this latter capacity, P describes the situation of reference under those features that make it eligible as a member of {sit1}, with what that entails in terms of expected course of action).
- The other consists in setting this categorizing aspect aside to allow attention to be redirected to Pc as an intentional content.

Once (8) has been perceived to be expressing an argumentative relationship between the two states of affairs (i.e. Ludwig feeling neglected, and the remedial action expected of H), the shift upwards makes it possible for (8) to be interpreted as marking a belief that the first state of affair – as categorized – is conducive to the second state of affairs. As a consequence, the act performed via the utterance of (8) is reinterpreted as a commitment on S's part to Pc being an appropriate categorization for the first state of affairs. (As will be recalled, this commitment provides the basis for the manipulative phase: by thus committing herself to Pc, S is attempting to get H to endorse her categorization of the situation. Inasmuch as this categorization entails a further categorization of the situation, and the linkage between the corresponding categories of situations is part of the shared background, S's attempt amounts to an attempt to get H to accept this further categorization.)

The above scenario does not exclude that (8) might first be interpreted as marking a commitment to the state of affairs in P being verified. S's commitment to Pc being an appropriate categorization of the situation of reference would arise as a later interpretation, once contextual and/or co-textual input has been brought to bear. In this scenario, as in the previous one, reaching an argumentative interpretation would be dependent on viewing P under two further aspects in addition to the initial one. Individual interlocutors in different circumstances may proceed differently in their attempt to identify an argumentative intent; nonetheless, with the discovery of the switch in perception involved in viewing P under a different aspect, a more direct access to this higher-level meaning becomes available.

Before turning to the third criterion, we need to ask ourselves where P's categorizing aspect comes from. The answer conceivably lies with what Searle (2001, pp.158–60) calls the 'generality requirement':

> When I say, 'That is a man', I am committed to the claim that any entity exactly like that in the relevant respects is also correctly described as 'a man' [...] [this] generality is built into the structure of the language itself, and indeed when it comes to the application of language, it looks as if we get ought's from is's wherever we turn. From the fact that an object is truly described as 'a man', it follows that you ought to accept relevantly similar objects as men. It is impossible to use language without those commitments.

In other words, when I make an assertion, whose point is to commit myself to P being the case, in virtue of my commitment in a public language, I am also committed – though this is not the point of the act – to the generality requirement, that is, to the fact that similar states of affairs would be described as P. What counts as a certain type of state of affairs can be a matter of social and cultural conventions, so, in addition to a commitment in a public language, one should probably also invoke a commitment in a shared local background knowledge. Whatever the case may be, in light of what precedes, there is nothing mysterious about the fact that P should have a categorizing aspect, whereby it implicitly describes a category whose members are similar to the situation of reference. In the case of acts of argumentation, this categorizing aspect comes into focus in the case of acts of argument-ation, through the mediation of an attentional shift from P as a description of a certain state of affairs, to P as a category judgment about the situation in question. To be more accurate, this shift is from P's representational capacity to its categorizing capacity, to cater for the fact that we are dealing with a category judgment concerning a specific situation, rather than a general description of a category of situations. That P should now categorize a specific situation follows from the fact that S's commitment to P (in a categorizing capacity) is no longer implicit, but has now become the point of the act, with what this entails in terms of possible reference to a particular situation. For future reference, one may want to note that the generality requirement has another aspect to it:

> if I am committed to recognizing similar instances as cases of men, my commitment in a public language requires that I think other people ought also to recognize this as similar cases of men.

The third criterion, which is about the directionality of the shift, is also satisfied. Indeed, the shift is from a domain of reference which is closer to the external world – one comprising types of situations, and, more to the point, relationships between them – to one that is less so, namely a subjective world consisting of the same entities, but seen from a subjective angle of viewing (in the event, an intentional one).

Inasmuch as acts of argumentation are speech acts, it is hardly surprising that the target domain should be the above subjective world. What might require further clarification is why the initial domain is not the external world, but an intermediate one. This follows from the fact that P's categorizing aspect is the one involved in the interface with the higher level. To say that P expresses a category judgment of the situation of reference is to say that the latter is perceived from the perspective of a shared social construct (in the form of a complex category), whereby one set of situations is linked to another. Alternatively, one may say that the object of perception is the relationship between the situation of reference and another type of situation, with this relationship being determined by a shared construct. From the angle of viewing provided by this construct the situation of reference is seen in terms of what it entails with respect to a future state of affairs. In other words, we are dealing with an intermediate shift, from the external world to an inter-subjective world, where the relevance of states of affairs lies in their being the basis for future action.

But why do we have an extrinsic angle of viewing (which is what the social construct provides)? Insofar as the connection between one state of affairs and another is not a matter of logic, but arises from a collective experience of them, the shift from P in a representational capacity to P in a categorizing capacity is not a simple question of shifting the window of attention: there has to be an (extrinsic) angle of viewing arising from a set of assumptions about what constitutes a basis for taking one course of action rather than another.

The proposed (general) directionality for the above shift is consistent with the kind of semantic layering found in utterances expressing an argument. When faced with (8), one can be more confident about the content of the propositional meaning than of the nature of the speech act. From S's standpoint, the propositional meaning level can also be seen to be prior to the speech act level and to be better established in terms of encoding. Thus P cannot function as an argument unless it is assumed that the state of affairs it describes is verified, which is consistent with how denying that the latter is the case amounts to invalidating the argument. Differences in degrees of encoding are manifest in the following contrast: while S cannot, without inconsistency, deny her commitment to the representational function of the propositional meaning, no such difficulty arises with respect to the act of argumentation. Consider (8) again: while S can hardly deny that P is meant to represent a state of affairs (whether we are dealing with a true representation is beside the point), she would not be similarly bound by a commitment to an argumentative interpretation of the act, as attested by the acceptability of:

(12) Ludwig feels neglected, but I am not saying anyone should attempt to remedy the situation.

(Where S is opposing the implication that (8) was intended as an argument.)

What was said above regarding the intentional content of the belief underlying acts of argumentation is consistent with the presence of an intermediate layer, namely one concerned with P in a categorizing capacity. Support for this hypothesis can be found in the following observations:

(a) P, as a representation, can be dissociated from P as a category judgment. Thus, in response to A's :

(13) A: I would go and see a doctor if I were you: You are not getting better.

B can reply:

(14) B: OK, I am not getting better, but I am too busy.

The overall effect of (14) is that B dismisses the idea that the situation calls for a visit to the doctor on grounds that he is too busy. What interests us here is the stance being taken towards P: in saying 'OK, I am not getting better', thereby acknowledging P as a true representation, B (in light of what follows) implies that P is being brushed aside as an appropriate assessment of the membership status of the situation. P, one may object, is still viewed by B as a categorization of the situation – albeit one he does not retain – and isn't the opposition introduced by 'but' one between two alternative ways of categorizing the situation? This objection overlooks a critical difference: the opposition is not concerned with the two category judgments as such, but with what they entail in terms of future action on B's part. What matters is whether P under present circumstances constitutes an appropriate way of viewing the situation, that is, one conducive to a certain type of action. On that basis, whether P in other circumstances can be an appropriate categorization is immaterial from B's standpoint. In other words, the dissociation that counts is the one between P as representation, and P as an appropriate categorization (as opposed to a possible categorization).

(b) In a context where P has just been put forward as an argument, an unqualified explicit commitment to P being the case constitutes a commitment whose object is primarily P as an argument, or – in our terms – as a categorization of the situation of reference. Consider this alternative response to (13):

(15) B: I am afraid you are right.

A natural interpretation for (15) is that P is being used by B as an argument, or a reason to go and see the doctor, rather than in a representational capacity.

On the basis of B's reply, his interlocutor would be entitled to assume that a visit to the doctor is on the cards, and (16) would be an appropriate response on his part:

(16) A: Oh good! I am sure that they'll find a slot for you tomorrow.

By contrast, (17), which only presupposes S's commitment to P as a true representation, would sound odd:

(17) A: Oh good! But do you propose to see a doctor?

In terms of constraints on continuation arising from (15), a similar result can be obtained. Thus, while a continuation based on P as a category judgment constitutes a normal development of the same discourse intention, one based on P, as a representation, would mark a change of intention. An example of this contrast is provided by (18) and (19), as alternative continuations to (15):

(15) B: I am afraid you are right.
(18) B: So I suppose I would do well to make an appointment with the doctor.
(19) B: ? Which means that the normal expectation is that I should be better by now.

(c) In the same situation as in (13), P cannot function as an argument if it is being rejected as a true representation, as attested by (20), an alternative response to (13):

(13) A: I would go and see a doctor it I were you: you are not getting better.
(20) B: On the contrary! I am feeling a lot better.

On the (strict) basis of (20), A can only infer that B does not intend to seek medical attention, and (21), which carries this assumption, would make a perfectly natural response to it; (22), which implies the opposite, is inappropriate:

(21) A: That's your back!
(22) A: I am sure they can fit you in as early as next week.

(d) Rejecting (13) on grounds that P is not verified (cf. (20)) amounts to an outright polemical act, which confirms its status as a presupposition. By contrast, on the basis of (14) – which preserves P as a true representation and rejects it as the only possible way of categorizing the situation – the interlocutor, arguably, does not have as much grounds for complaint.

Together, the above observations support the postulation of an intermediate level, one concerned with categorization and interfacing with the higher level. Point (a) shows that we are dealing with two distinct levels; point (b) identifies the categorizing level as new information and the source of constraints on continuation; points (c) and (d) confirm the implication that the representational level is presupposed.

To be regarded as favourable grounds for divergent thinking, acts of argumentation are expected to meet a fourth and final criterion: there should be a processing problem for which the use of divergent thinking under consideration would provide a more efficient solution.

In view of our claim that argumentational meaning requires a different application of divergent thinking, this problem should be distinct from the one common to all speech acts. In the case of speech acts in general, the problem is one of identifying S's illocutionary intention on the basis of a propositional meaning; and H's ability to view that propositional meaning as the intentional content of a mental state is what provides direct access to S's illocutionary intention, that is, her intention to commit herself or H with regard to the direction of fit between words and world. In other words, this intention is one of either committing herself to P being a true representation, or to place H under an obligation to bring about a state of affairs that will satisfy a desire she expresses by means of P. Since acts of argumentation are speech acts, the problem is, again, one of identifying S's illocutionary intent in expressing the propositional meaning. However, as will be recalled, we are dealing with a different intentional content. In saying:

(8) Ludwig feels neglected.

with an argumentative intent, S purports to place H under an obligation to construe the situation of reference (i.e., a certain behaviour on the part of Ludwig) as falling under a certain category of situation ({sit1}), one calling for a certain state of affairs to be brought about, which exhibits the membership criteria of another category of situation ({sit2}), in virtue of a pre-existing link between {sit1} and {sit2}. In short, while the same attentional shift – from P in a representational capacity to P as an intentional content – would give us the nature of the expressed mental state and the general illocutionary intent, in order to obtain the correct intentional content for the mental state and the correct specification for the illocutionary intent, there is a need for a second[16] attentional shift, one concerned with viewing P under a categorizing aspect. In interpretative terms, this amounts to shifting the focus from one issue to another, from whether (8) represents an actual state of affairs, to whether (8) provides a possible argument for a certain conclusion. To put it differently, the processing problem in the case of acts of argumentation involves a further complication relating to the identification of the conditions of satisfaction of the act. However, S's intention is not merely that H should construe P as a possible argument, but – more to the point – that H should use P as an argument to reach a specific conclusion. The extra complication thus lies in the identification of background assumptions susceptible of providing further instantiation to the conditions of satisfaction. Identifying these background assumptions – in the event, complex categories – can be made easier if the co-text provides explicit conclusions (e.g. 'You must do something about it'). Either way, the identification of the conditions of satisfaction may be said to involve that of a sub-intention concerned with the means made available to H, in the form of an imposed complex category. By validating the identification and acceptance of Q on the basis of P, the relevant complex category speeds up the fulfillment of S's conditions of satisfaction. So the processing problem to which this further use of divergent thinking provides a shortcut is that of gaining direct access to the relevant complex categories in our store of categorized factual knowledge. As for the connection between this instance of context constructing and the change of aspect affecting P, it is as follows: viewing P in a categorizing capacity presupposes a perspective from which P is perceived, and one provided by a complex category to which this perception constitutes an implicit reference. Thus, if in addition to what precedes it can be further assumed that complex categories of the type under consideration

16 Why 'second' shift is because it derives its significance from its role vis-à-vis the higher-level mental state, an illocutionary intent. If it were not for the latter, the aspect under which P is viewed would not be an issue. Now, from the interpretative standpoint, things may not be so clear-cut. The 'second' shift could be regarded as prior, since the ability to view P in a categorizing capacity underlies the shortcut needed to gain quick access to S's argumentative intent.

pertain to an inter-subjective domain, then viewing P in a categorizing capacity should – to a lesser or greater extent – trigger this kind of context constructing.

What precedes prompts the following question: how does categorization relate to the point of the act? The answer is that the category, {sit1}, being referred to in connection with P is linked to a further category, {sit2}. Thus the perspective of {sit1}, from which the state of affairs in P is envisaged, is more accurately that of {sit1} as seen from the perspective of {sit2}, or that of its relationship to {sit2}. Accordingly, in categorizing the state of affairs in P, the point of reference H is expected to use is not {sit1}as such, but {sit1}as conducive to {sit2}, which determines the conclusion H must endorse in fulfillment of the conditions of satisfaction of the argumentative act.

Insofar as the expression of this conclusion provides a continuation to that of the argument, from a textual standpoint, the problem of identifying the context constructing sub-intention coincides with that of identifying the appropriate type of continuation or response (as the case may be). So, in terms of this level of analysis, the processing problem this further use of divergent thinking is meant to cater for may be construed as one of anticipating the upcoming text.

By way of closing this section: the outcome of the above discussion is that acts of argumentation not only satisfy the four criteria relating to divergent thinking, but the way they meet the last three, sets them apart from bona fide speech acts, whose emergence we also attribute to divergent thinking. Skipping over Criterion 1 then, the difference exhibited by acts of argumentation in the way they satisfy Criteria 2, 3 and 4 is as follows:

- The aspect under which P interfaces with the higher level is concerned with categorization, rather than representation (= criterion 2).
- The directionality of the shift, though involving the same target domain as in the case of other speech acts (namely a subjective domain, determined by an intentional point of view) takes as its initial domain one that is further away from the external world than what we had in the other case. This initial domain, referred

to as 'intermediate', may be construed as the domain of discourse (as a sub-domain to the external world), but seen from the angle of viewing provided by social constructs (or complex categories pertaining to a social and cultural background). (This is what is meant by 'viewing P as providing a category judgment of the situation of reference', as opposed to providing a mere representation of it.)

- The processing problem for which divergent thinking provides a shortcut is that of identifying a sub-intention concerned with constructing the appropriate processing context. Inasmuch as this context is evaluative, this processing problem can be further construed as one of identifying on what grounds H is expected to fulfill the conditions of satisfaction of the act, that is, accept a certain conclusion along with the attendant continuation that expresses it. These grounds consist of complex categories that validate a certain linguistic response on the basis of a given communicative situation.

Further arguments in support for divergent thinking

In light of what precedes, acts of argumentation present all the conditions that would make them susceptible to a further application of divergent thinking. Now, if divergent thinking is responsible for the emergence of argumentational meaning, in the sense of being involved in the encoding of relevant constraints, those contents that provide the input to the encoding process can be expected to bear the hallmarks of this ability.

Leaving aside, for the moment, the details of these hallmarks, let us begin by setting out the main assumptions we shall be needing, some of which have already been introduced:

(1) What is meant by 'content' is germane to Kosslyn and Koenig's 'representation'[17] (cf. Chapter 2, note 13). For current purposes,

17 With respect to 'representation', a cautionary remark is in order. Though widespread in cognitive science, the use of 'representation' is not without raising

however, one should also bear in mind that the term subsumes both 'intentional states' and 'preintentional assumptions, stances, practices' (of which more later), and includes procedural features.

(2) Encoding, or the process whereby a new association between two contents is stored into memory (Kosslyn and Koenig 1995, pp.355–6), is a feature of automatization. As a prevailing means of achieving processing efficiency, this process is carried out, among other things, by a strengthening of frequently used pathways, those linking contents which are essential to reaching a certain outcome. In the case at hand, these contents are the ones on which the retrieval of the conditions of satisfaction of S's argumentative intent is dependent.

(3) Divergent thinking has a crucial role to play in this routinization whose end result is the type of encoding that interests us: as a mnemonic strategy it imposes its own requirement on the pathways involved, giving rise to a shortcut that speeds up the interpretative process. (In view of the fact that we are now dealing with two distinct applications of divergent thinking, one 'illocutionary', the other, 'aspectual', more will need to be said about the role each plays vis-à-vis this shortcut).

If divergent thinking contributes to processing efficiency in the manner just described, then, the contents involved in the linkage it provides can be expected:

(a) to have a key role in the accelerated process whereby the relevant interpretative outcome is reached (i.e. the conditions of satis-

certain problems. Thus, as Edelman (1989, p.13) points out:; 'to talk about "representations" in the brain is to make certain hidden assumptions about how a conscious observer catalogs what is assumptively described as "information" in the macroscopic physical order, for example, the categories of things and how they are arranged.' Searle (1992, pp.212–14) discusses this 'homunculus fallacy' at greater length. Short of coining a new term, which may not be desirable, I will opt for 'content', which seems to me less problematic.

faction of the higher-level mental state, of which the argumentative act is the expression);

(b) to stand in a relation that is consistent with the type of attentional shift entailed by divergent thinking.

Our concern with the set of contents linked up by divergent thinking stems from the assumption that the encoding of one or more of these contents is responsible for the constraints under consideration.

In what follows, our main task is to identify plausible candidates for encoding. Once the relevant set of candidates has been identified, we will need to establish to what extent they coincide with those contents that bear the hallmarks of divergent thinking. Where the latter derive from is something that will have to be clarified beforehand.

Before we set out to address the above points, we may want to bear the following in mind: in considering acts of argumentation one must be careful to distinguish between two types of constraint on continuation, one common to all illocutionary acts, the other, specific to argumentational meaning. The former (henceforth, 'illocutionary') concerns the type of response that would fulfil the conditions of satisfaction of the act; the latter (henceforth, 'argumentational') determines the type of content that should instantiate the above response. Consider:

(23) Does Octavio have an appointment with Dr Lee?

The illocutionary constraint – that the appropriate answer should be a confirmation or a rejection of the propositional content – can be directly linked to the conditions of satisfaction of the act: once (23) has been identified as marking a question, what constitutes an appropriate response becomes immediately clear on the basis of its conditions of satisfaction. As a type of illocutionary act, acts of argumentation would, as a matter of course, possess an illocutionary constraint, which derives from their own conditions of satisfaction. Consider:

(24) He finds it challenging.

in response to 'Do you think X will do it?', where 'it' refers to an assignment of some kind. If (24) is taken to express the premise of an act of argumentation, with what that entails in terms of conditions of satisfaction, the constraint on response, will, accordingly, be that H should accept a certain conclusion on the basis of the premise put forward. Now, alongside an illocutionary constraint, acts of argumentation also have an argumentational constraint, though this is not invariably the case. Consider (24) again: the fact that S describes the assignment as 'challenging' creates the expectation that the conclusion will be something to the effect that the person in question will, at the very least, have the motivation to consider undertaking the assignment. Had the latter been described as 'daunting', the expectation would have been reversed, with the appropriate conclusion expressing a view ranging from a mere reluctance to a clear refusal, on X's part, to undertake the assignment. So, unlike illocutionary constraints, which are determined by the structure of intentionality, argumentational constraints – at least those under study – appear to arise from the conceptual structure of individual lexical items. Our concern, in what follows, will be exclusively with constraints of the latter type.

But first, we need to remind ourselves of what argumentational constraints are about. The following examples (from Anscombre, 1995) were previously quoted as (1) and (2):

(25) a. Pierre est riche.
Pierre is wealthy.
b. Il peut s'offrir n'importe quoi.
He can afford anything he fancies.
c. Il est donc avare.
He is therefore stingy.

(26) a. Marie est belle.
Marie is beautiful.
b. Elle séduit tous les hommes.
Men all fall for her.
c. Elle est donc coquette.
She is therefore a flirt.

In both sets of examples the existence of a constraint is manifest in the fact that (a) has (b) as a preferred continuation, in the sense that the latter is what most readily comes to mind in the absence of further contextual information. The same cannot be said of (c), which could not be predicted on the basis of (a) alone, but requires additional contextual assumptions.

AWL's view on the matter, as will be recalled, is that preferred continuations of the above kind derive from the semantic value of certain lexical items, which are construed in terms of intrinsic topoï. Where on an objectivist construal the meaning of a lexical item is defined in terms of the invariants traits of a given object, in AWL this meaning consists of a set of situations associated with the object in question. Our take of the matter was that, predictable as it may be (on the basis of the semantic value of the lexical item occurring in the preceding utterance) a preferred continuation is not an explicitation of some aspect or property of the object being referred to; rather, it is the representation of stereotypical situations featuring abilities, stances and practices associated with it (to use Searle's terminology). (25) exemplifies a linkage between a property (being wealthy) and the kind of practice those who possess it can engage in; (26) is a case of another property (beauty) being associated, this time, with a certain ability (that of attracting people). (4), repeated below as (27), is a case of an object (a problem) being associated with a certain expectation regarding the amount of effort required in dealing with it:

(27) a. Il y a un problème.
There is a problem.
b. Mais il est facile à résoudre.
But one that can easily be solved.

c. Mais il est difficile à résoudre.
   But one difficult to solve.

On the basis of what precedes, one would expect the contents involved in the encoding of argumentational meaning, to bear a strong resemblance to topoï, or complex categories, in the sense of presenting, in part or in full, a structure consisting of two associated contents, one corresponding to an object or a state of affairs featuring it, the other, to relevant stances, abilities, or practices.

Setting aside, for the time being, the problem of whether, or to what extent such a structure is susceptible to encoding, I propose that we turn to our next concern, which is to determine whether such a content can plausibly be found among those bearing the hallmarks of divergent thinking. As a preliminary to addressing this question, we must, as announced, spend some time explaining the source of this latter set.

Inasmuch as the contents making up that set are the ones whose occurrence in the context of utterance would ensure the correct identification of the conditions of satisfaction of the act, they would be found among the wider set of SN conditions characteristic of argumentative acts. These conditions – whose presence in argumentative situations help steer H towards a certain conclusion – are construable in terms of intentional states[18] and different varieties of (preintentional) know-how S possesses. Like all SN conditions they pertain to an idealized situation, rather than to actual ones: the latter, indeed, cannot be expected to exhibit the full range of nonaccidental contextual features necessary to the successful identification of the act, any more than they can be expected to exclude accidental features.

18 The difference between intentional states and preintentionnal know-how is not dissimilar to that between explicit memories, which are accessible to introspection and available for other purposes and implicit memories, which are not, and can only be activated in the performance of skills.

SN conditions

Following the general layout proposed by Searle (1969) for SN conditions in the case of illocutionary acts, an idealized argumentative situation can be described as one in which:

(1) S believes that P is a true description of the situation of reference (SR).
(2) S assumes that H would also hold the belief that P is a true description of SR, or at least ought to.
(3) S assumes that P is an appropriate characterization of the membership criteria of a whole set of situations. This set of situation is a simple category (e.g. {sitp}), if P is envisaged as a lower-level unit of meaning. It is a complex category ({sitp}<=>{sitq}), if P is envisaged as [P<=>X], that is, a discourse-level unit of meaning).
(4) S assumes that H holds similar views in respect of P, or at least ought to.
(5) S possesses the assumption that there is a complex category {sit1}<=>{sit2} such that:
    (i) members of {sit1} are describable as P, and those of {sit2} are describable as Q;
    (ii) {sit1}is linked to {sit2}, in the sense that members of {sit1} would normally evoke those that are members of {sit2}, and vice versa.
(6) S assumes that H shares her assumption in (5).
(7) S believes that P appropriately categorizes SR as a member of {sit1}.
(8) S assumes that SR', the situation evoked by SR, would be appropriately categorized by Q, on the basis of (3), (5) and (7).
(9) It is not obvious to both S and H that, in the normal course of events, H would come to view P as an appropriate categorization for SR.
(10) Nor is it obvious to both that, in the normal course of events, H would be committed to Q as an appropriate categorization of SR', on the basis of viewing P as an appropriate categorization for SR.

(11) S wants H to commit himself to P as an appropriate categorization of SR.
(12) S wants H to commit himself to Q as an appropriate categorization for SR' on the basis of his commitment to P as an appropriate categorization for SR.
(13) In producing P, S undertakes a commitment to the effect that P is an appropriate categorization for SR, which amounts to a commitment that Q should follow from this prior commitment, in virtue of her above commitment to the complex of relations involving {sit1} and {sit2}, P and Q, P and {sit1}, and Q and {sit2}.
(14) S's undertaking in (13) counts as an attempt to get H to recognize P as an appropriate categorization for SR by way of getting him to accept that Q follows from P, in virtue of H's presumed commitment to the complex of relations involving {sit1} and {sit2}, P and Q, P and {sit1}, and Q and {sit2}.

A few comments are in order:

- S's intention in making an act of argumentation – whose point is to place H under an obligation to accept that Q follows from P – can only be the intention it is as part of what Searle (1979, 1983, 1995) calls a 'Network of intentional states', which includes a desire on S's part that H accepts Q on the basis of P, and subsidiary intentional states concerned with the means to achieve the above goal. This intentional network shades into, or, rather, is permeated by the Background, or set of preintentional abilities, stances, assumptions, which form the preconditions that enable intentional states to function, and all representing, in general, to take place. There is no clear dividing line between the Network and the Background, and what I have listed as assumptions should probably, for the most part, be qualified as 'preintentional', since, to use Searle's criterion, they are not part of the semantic content of the components.[19] Among the assumptions listed, those that seem clearly preintentional are:

- (9), (10), (13) and (14), which concern a certain know-how about speech acts S and H can be expected to possess, as normal human beings living in society;
- and (5), which are about P's categorizing capacity.

Note that while (9) to (15) belong to a deeper Background, (3) and (5) should probably be attributed to a more local Background, since what counts as an appropriate character-ization of a category of situations and determines what categories of situations are linked, is, at least in part, a matter of cultural background. (2), (4), (6), and (8) I take to be intentional, that is, to be part of the intentional network, as they are directly relevant to S's intention. What I have listed as beliefs and desires of S is part of the Network:

- In light of what has been said regarding the nature of the Background and the way it functions, the issue does not arise as to whether the proposed list can be expected to be exhaustive. Furthermore the various contents it features must be seen as exhibiting varying degrees of intentionality.

Turning now to the raison d'être of some of these contents:

- (2), S's assumption that H takes P to be a true description of SR follows from the fact that S's belief in (1) is presupposed.
- S's assumption in (3) – that P appropriately characterizes the membership criteria of a whole set of situations – follows from the generality requirement. This requirement, to jog our memory, is one S must observe in virtue of her commitment to the fact that

19 This point is one we need to come back to. As we saw in the previous chapter, it is apparently possible for background contextual assumptions to make their way into the semantic structure of the utterance, or rather become associated with their motor representations, as opposed to the semantic contents of individual words

situations similar to a given SR in relevant respects (i.e. {sit1}) would be described as P (= (a)). This commitment to (a) entails another commitment, namely that P appropriately describes the same whole category of situations. Insofar as S's commitment to the generality requirement also involves a commitment to thinking that other people ought to recognize that the same category of situations would be described as P (= (b)), this amounts to a further commitment, namely to thinking that other people ought also to recognize that P is an appropriate description of that same category.

- (3) is neutral with regard to the status of P. If P is envisaged as a lower-level unit of meaning, the set of situations it characterizes is a simple category ({sitp}). On the other hand, if P is construed as part of a sequential or discourse-level unit of meaning (i.e. as [P<=>X]), then the category in question should, accordingly, be a complex one, for example, {sitp}<=>{sitq}, where {sitq}, the category of situations characterized by X (and later to be instantiated as Q), is linked to {sitp}, in the sense that situations that are members of {sitp} would normally evoke those that are members of {sitq} and vice versa. The two corresponding versions of (3) will be referred to as (3a) and (3b).
- The rationale behind (5) and (6) – which concern S's assumption regarding the existence of a complex category {sit1} and {sit2}, and her further assumption that H shares the same assumption – may be laid out as follows: just as communication cannot take place unless S and H share the same linguistic conventions and the same background knowledge, so S cannot engage in an argumentative act with any reasonable expectation of success, unless she can safely assume that both interlocutors share a certain knowledge about 'how the world works […] and a set of abilities for coping with the world' (Searle 1995, p.131), including a certain cultural know-how about what domains of experience would be associated, what courses of action, given a certain type of situations, would normally be called for, and what being a certain type of person entails in terms of stereotypical behaviour (e.g. being oppressed and seeking liberation, being pro animal rights and favouring humane mouse-traps), and what

stances and practices are normally expected with regard to given objects.

Relevant subset

Having established a list of essential recurrent features such as would be found in an idealized argumentative situation, our next task is to identify the relevant subset in relation to divergent thinking.

Of the fourteen SN conditions listed, the most plausible in that respect are arguably:

- (3), S's assumption that P is an appropriate characterization of the membership criteria of a whole set of situations;
- (4), S's assumption that H holds similar views in respect of P, or at least ought to;
- (5), S's assumption regarding the existence of linked categories of situations, or complex categories of the form ({sit1}<=>{sit2});
- (6), S's assumption that H also shares her assumption in (5);
- (7), S's belief that P appropriately categorizes SR as a member of a category {sit1};
- (8), S's assumption that SR' would be appropriately categorized as Q.

All the above are preconditions that must be met if the obligation imposed on H is to be validated. Unlike what happens with a communicative act, where S's commitment to P – albeit in a representational capacity – is the intended cause of the obligation, in this case, in addition to S's commitment to P (in a categorizing capacity) the obligation also requires that S and H share the same assumptions about what constitutes the membership criteria of certain categories of situations, and what categories of situations are linked (in virtue of the fact that the domains of experience they represent are connected as part of a common culture).

As our next task, we must determine which among the contents in the above subset bear the hallmarks of divergent thinking. To refer back to conditions (a) and (b) set out earlier, the relevant contents are the ones whose linkage should provide a plausible shortcut to a certain interpretative outcome; furthermore the relation in which they stand to one another should be construable in terms of what divergent thinking entails as a process.

Divergent thinking – in one of its alternate characterizations – is the ability that 'helps us inhibit the tendency to act on simple correlative stimulus relationships and guides our sampling of higher-order sequential or hierarchic associations' (Deacon 1997, p.265). So, if divergent thinking is involved in the interpretation of an argumentative utterance P (expressing a premise in support of a conclusion Q), the shortcut it provides should:

(a) inhibit the tendency to view P in a representational relation to SR, thereby shifting the focus to P in a categorizing capacity, or, at least, creating the possibility for this to occur;
(b) shift attention to higher-order sequential associations, that is, to P in its relation to an intended continuation, or as a constituent of a higher-order unit of meaning.

Whether these two processes are simultaneous or sequential, our assumption is that the outcome of the former is available to the latter.

Schematically, (a) and (b) could be represented as in Figure 1 and Figure 2, respectively:

Figure 1

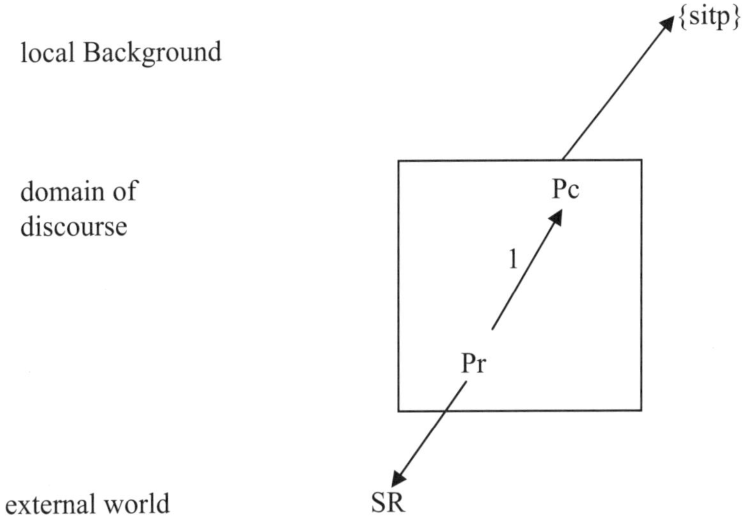

During the inhibitive phase (Figure 1), the focus of attention (arrows #1) shifts away from P in a representational capacity (= Pr) – as exemplified in its relation to SR – thereby making it possible for another aspect of P to come into focus, in the event, its categorizing capacity (= Pc). Viewing P under this latter aspect entails a reference to a category situation {sitp}. The square box surrounding Pr and Pc marks P out as a lower-level unit of meaning. Domains of experience (local Background, domain of discourse, external world) are indicated vertically on the left hand side.

Figure 2

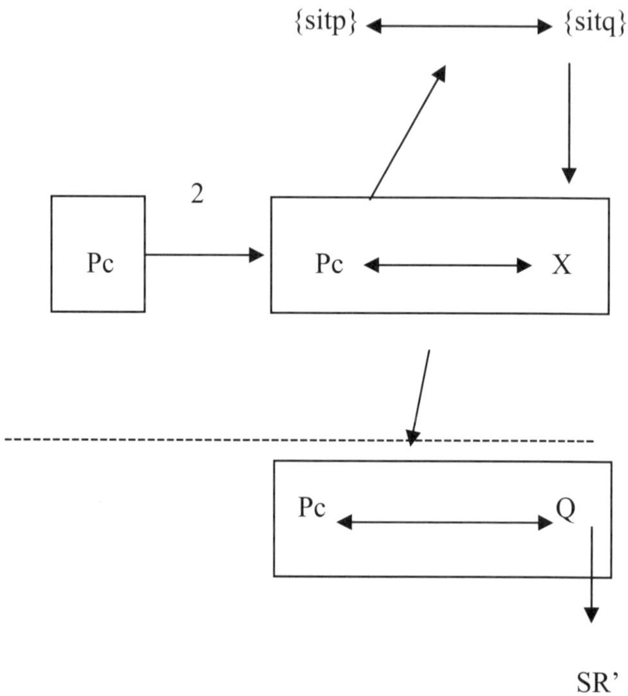

The shift to higher-order associations (arrow #2) corresponds to a construal of Pc whereby it is viewed as part of a sequential unit of meaning (top rectangular box), where the continuation it is linked to is as yet unspecified by the interpretative process (hence the X). The double-headed arrow between P and X indicates a reciprocal relation, in virtue of which either can be seen from the vantage provided by the other, or further defined in terms of its link to the other. The extent that P's categorizing aspect is carried over to the higher level, it, accordingly, also entails a reference to a category, but a complex one, this time, namely, {sitp}<=>{sitq}, where {sitq}, if correctly selected among possible categories that might be linked to {sit1}, will provide Q (i.e. the continuation to P). That P's categorizing aspect is carried over to the higher level is compatible with the two types of relationship envisaged for the upwards shift and the aspectual switch. The bottom rectangular box represents a fully instantiated version of the top one. It is, strictly speaking, not concerned with the shift, hence the dotted line separating the two versions. Again the numbered arrow stands for what divergent thinking does. The single-headed unnumbered arrows represent preexisting relations (upwards), or interpretative outcomes (downwards).

In terms of the decision making process, there is an obvious parallel between the aspectual switch and the reference to complex categories, on the one hand, and the categorization of the initial situation and the process whereby the store of complex categories is accessed, on the other. The intermediate stages in the interpretative version reflect the attentional shift required to deal with the fact that the lower level structures that serve to express higher-level meanings are conventionally associated with lower-level meanings.

In light of what precedes, which contents would be directly involved with divergent thinking, in the sense of providing the shortcut through their linkage?

But first, let us remind ourselves what interpretative outcome this shortcut is meant to provide a quick access to. Based on what was said earlier, from an illocutionary standpoint, what this application of divergent thinking does amounts to providing S's mental state (of which the argumentation act is the expression) with the appropriate intentional content: had it not been for this application, P as a representation of SR would have gone on to instantiate that intentional content. Instead, this intentional content is now that Q should be the appropriate categorization for SR', on the basis of P being the appropriate categorization for SR. In terms of SN conditions, this points to (7) and (8). So, given that (7) and (8) coincide with the relevant interpretative outcome, the contents whose linkage would provide the appropriate shortcut should be:

- (1), S's belief that P is a true representation of SR.
- (3), in either versions of (3a) or (3b) (with (3a) = S's assumption that P, as a lower level unit of meaning, is an appropriate characterization of a set of situations {sitp}; and (3b) = S's assumption that P, as a sequential unit of meaning (or [P<=>X])) is an appropriate characterization for a complex category of the form {sitp}<=>{sitq}.
- (5), S's assumption concerning the existence of a complex category {sit1}<=>{sit2}.

The selection of the above set calls for some comments:

- Why (1), which was omitted from the previous subset, should find its way here has to do with the fact these contents are viewed from a procedural perspective: Insofar as processes require a starting point as well as an endpoint, the attentional shift under consideration would have to have a prior perceptual object, from which the focus would have moved away. This prior object is P under a representational aspect.
- The linkage of (1) to (3a) and (3b) corresponds to two shifts occurring in a sequence: one from egocentric to allocentric associations (i.e. from P in a representational capacity to P in a categorizing capacity); the other, from one construal of P to another, that is, from P as a lower level unit of meaning to P as a constituent of a higher level unit, [P<=>X].
- If the two shifts are simultaneous, (1) would be directly linked to (3b).
- A question that arises is: why should the focus of attention be redirected to P in a categorizing capacity, once it has shifted away from P in a representational capacity? Apart from the trivial reason that no other alternative might be available, one that readily comes to mind is that there is a connection between the directionality of the shift – which is upwards – and viewing P in a categorizing capacity. P's categorizing relation vis-à-vis SR may be seen as a type of representational relation, albeit one which is mediated by reference to a category of situations. Now referring to such a category necessarily involves an attentional shift to a domain of experience that is further away from the external world. The upshot of this is: if at the onset of the process a shift upwards is already in the offing, the chances are that this will influence which aspect gets selected.
- The fact that (3a) and (3b), which concern P's categorizing aspect, come into focus as a result of the aspectual switch means that from background assumptions they find themselves raised to an intentional status, thereby becoming an integral part of the Network. Note that this does not apply to (5), despite its close connection to (3), although this entails that it has moved closer to the 'perceptual field'. The reason why it was listed is because it constitutes the endpoint of the new direction of fit brought about

by the switch to P's categorizing capacity: from words-to-world, this direction of fit has become from words-to-mind, with 'mind' being construed in the case at hand in terms of shared background knowledge.

Our next concern is whether (1), (3a), (3b) and (5), or alternatively (1), (3b) and (5) bear the hallmark of divergent thinking as defined by condition (b). In other words, is the relationship in which they stand accountable in terms of what divergent thinking entails as an attentional shift? On the basis of what was said above, it would appear to be the case. Thus, the link between (1) and (3a) is attributable to the inhibitive phase and would be brought about by the switch from one aspect of P to another, that is, from P in a representational capacity, to P in a categorizing capacity. The link between (3a) and (3b) corresponds to the upwards shift, whereby P is construed as part of a sequential unit of meaning, [P<=>X]. Now, if what we are considering is the link between (1) and (3b), this link, would, accordingly, be brought about by the conjoined action of the inhibitive phase and the upward shift. As for the link between (3b) and (5), it has to do with an aspect of the shift, namely its directionality, which has been described in terms of a change in domains of experience. (This link is not unlike a 'pointer' or representation that indexes another representation. Inasmuch as perceptual memories are not duplicated in associative memory, but the associations between them are – in the form of relations between token symbols – the latter (or rather the corresponding patterns of activation) would be associated with pointers pointing back to perceptual representations, thereby allowing the references of the symbols to be 'unpacked') (Kosslyn and Koenig 1995, p.358).

For an illustration of how this shortcut is meant to work, consider:

(28) The cat is allergic to dust (= P).

(as said to a house-sitter (H), in whose care the cat is to be entrusted.) Assuming that the context leaves open the possibility of two alternative interpretations, one non-argumentative (e.g. S is simply informing H of one of his future housemate's peculiarities), the other, argumentative (e.g. S expects H to draw from P a conclusion Q regarding the necessity of engaging in house cleaning on a regular basis), this is how having a shortcut (arising from the 'aspectual' application of divergent thinking) would facilitate the retrieval of S's argumentative intent. Given (28) – whose default interpretation is that it (= P) describes a state of affairs (to which S is committed) – an argumentative interpretation, which normally should be dependent on contextual inputs, can become immediately available, if there is a routine in place. Such a routine entails that the focus of attention can simply be switched from one aspect of P to another (i.e., from P in a representational capacity, to P in a categorizing capacity). This allows P to be viewed as part of a sequential unit of meaning, where previously it was construed as a lower-level unit of meaning. P in its representational capacity is instantiated in this case by the fact that it describes a property of the cat; its categorizing capacity corresponds to the fact that in describing the situation of reference as a certain state of affairs, it implicitly presents it as a member of a category of situations it could also serve to describe. As for envisaging P as part of a sequential unit of meaning – rather than a lower-level unit – this corresponds to viewing 'The cat is allergic to dust', not as a purported representation of a state of affairs, susceptible of being true or false, but as a verified representation of such a state of affairs, whose point lies in what it evokes in terms of appropriate stances and courses of action.

Now, if no clear indications are available from the co-text concerning what course of action the state of affairs described calls for, a search process would be triggered, whose aim is to identify the relevant background assumptions about the category of situations referred to by P and their association with other categories of situations. These assumptions would be germane to what AWL means by 'extrinsic topoï'. This is in keeping with what happens during an identification process, which, as a matter of routine, accesses activity patterns encoded in memory in an attempt to categorize the stimuli under consideration. But even if the interpretative process can rely on this more general routine, the amount of work it has to do is still, arguably, considerable, unless there is an encoded link allowing a more direct access to relevant assumptions (in the form of complex categories involving situations featuring pets subject to allergies and scenarios of appropriate courses of action). As it happens, no such link appears to be available from (28): to say that someone is allergic to dust is not conducive to the conclusion that one should do something about it (even in the most compassionate of societies), at least not without further contextual inputs. Such a link, however, appears to exist in the cases discussed previously.

In what precedes we identified a set of contents whose linkage through divergent thinking would have provided the interpretative shortcut leading to the encoding of the constraints that concern us. The question we must now ask is: what is it that got encoded? If linguistic encoding is a matter of associating the representation of a symbolic structure with a semantic representation, then the question is about the nature of the linkage: are we dealing with a process that automatically accesses a specific complex category, given a lexical item, or with a search process which is constrained with regard to the type of knowledge to be accessed? Inasmuch as the constraints at hand are but weak constraints (since they give rise to preferred continuations) this point to the latter alternative.

So, given that we are dealing with a weak link, what would it be based on? The answer, I think, lies in dispositional representations. As will be recalled, both innate and acquired knowledge is embodied in dispositional representations (or 'sets of neuron firing dispositions within an ensemble of neurons'). Those dispositional representations that provide the basis for acquired knowledge not only contain the means to recall imageable knowledge 'used for movement, reason, planning, creativity, [but also] records of rules and strategies with which we operate on those images' (Damasio 1994, p.105). My proposal is that the type of dispositional representation that would be associated with the lexical items under consideration is of the latter kind, in that it contains the record of an interpretative strategy implemented by divergent thinking.

Chapter 4

Methodological issues

This closing chapter takes up some key issues prompted by the nature of this study. In attempting to view linguistic phenomena in terms of constraints arising from brain structure and function, the very first questions that come to mind concern the data.

Thus, if these data are to be envisaged from a neurocognitive perspective, there is a worry that the linguistic parameters that inform them will not allow for an adequate degree of interfacing with neurocognitive processes. Unless this condition is fulfilled, there is no reasonable hope of accounting for the former in terms of the latter.

As posed the problem could be misconstrued as a matter of clashing assumptions about language and the brain. In fact, the assumptions a linguistic theory may hold in that respect would appear – more often than not – to have far more implications for its own methodology than for the issue at hand. Thus, adherence to the idea of a strong language module can effectively free the linguistic model from constraints of a neurocognitive order, while still enabling it to regard cognitive aspects of language as legitimate objects of investigation. Opting for weak modularity, on the other hand, usually signals a readiness to seek a model and methodological principles in accordance with a more embodied view of language. A third possibility lies in steering clear of this issue altogether, as is the case with AWL, which does so on grounds of methodological integrity.

Far from being a matter of clashing assumptions, the above problem is rather one of finding an angle of viewing from which the data can plausibly be seen to arise from appropriate underlying processes. In the case of the meanings under study, this was made possible by the fact that they readily lend themselves to the exercise, as their emergence – which also happens to be the focus of our attention – was construable in procedural terms, in the event, in terms of language adaptation to processing requirements.

The next question is prompted by this shift in perspective: how does it impact on the nature of the data? In respect of the data at hand, an important change can be noted, which affects the individual status of each type of meaning in relation to the framework of reference. From a linguistic standpoint, the above meanings do not all fall into the same category: while certain illocutionary values and SN conditions are normally conferred a linguistic status – albeit to varying degrees – argumentational meaning tends to be regarded as non-linguistic (especially by non-practitioners of AWL). By contrast, these same meanings have a similar status from a procedural standpoint, even if the automatization processes they are presumed to emerge from may differ with respect to the level of completion they present.

A second consequence of this shift in perspective is that the susceptibility of these meanings to being recategorized as the same type of entity lends plausibility to the thesis that their respective emergence is part of the same phenomenon. From a linguistic standpoint, this was by no means obvious, since the data came from different frameworks. As part of the same phenomenon, these meanings become amenable to an account in similar terms.

This brings us to our next issue: in what terms is such an account to be given? Among the assumptions I took on board in selecting Deacon's co-evolutionary perspective was the following:

- Language, as a latecomer, can be expected to co-opt pre-existing means rather than evolve means of its own, especially from scratch.

This, however, does not tell us how does one go about deciding which means are appropriate (where 'appropriate' is taken to mean 'susceptible of providing the basis of a hypothesis consistent with known findings about brain structure and function')? Hence the need for this further assumption:

- Solutions that have proved to be of adaptive value in dealing with certain problems will normally be called upon again to address problems of the same type.

Given these premises, the next logical step is to characterize the problem at hand in as much detail as possible.

To the extent that it is safe to assume that without the emergence of those meanings interpretation would not have been fast enough to meet the needs of communication, the problem to which their emergence is presumed to bring an answer is best described as one of increasing processing efficiency. Quite apart from the fact that we do know – on the basis of our own experience – how much trickier interpretation is at the text level than it is at lower levels, this hypothesis is consistent with the model of language development advanced by Givón (1979). Givón, indeed, sees the rise of syntax as a response to increased communicative pressures brought about by an expansion of the range of communicative objects.

The problem having been categorized as one of boosting processing efficiency, we can turn our attention to those means that can plausibly be expected to subserve processing systems in general, and to cater for the specificity of the situation at hand, in particular.

The known way for processing systems to improve their efficiency is via automatization, a process whereby frequently used pathways become established. These pathways prove their usefulness in providing shortcuts between input and output points, where previously only more circuitous – and hence less adaptive – routes were available. (A normal decision-making process, indeed, requires much deliberation at each stage before an appropriate response can be selected and enacted.) The simplest and most direct form of automatization is exemplified by the stimulus-response model, which provides a direct link between perceptual and motor representations. However, meaning is known to be linguistically undetermined, a situation verging on the critical at the text level, where structures already associated with a propositional meaning (e.g. expressing a state of affairs in the world), are called upon to communicate a higher-order meaning (e.g. S's mental state), the expression of which serves to perform a speech act (e.g. a command). To compound matters further, interpretation at the level in question requires contextual intervention, as well as appeal to background knowledge. In other words, the sheer complexity and amount of the inferencing required would argue against the implementation of anything remotely like a straightforward stimulus-response type of connection.

If this model is to be ruled out, the tendency towards automatization would have to be implemented by different means.

As a preliminary to identifying those means, we must be clear about what types of constraint the interpretative process is subject to at the text level. Having these to hand, we could then proceed to find the best matches for them.

Given that at the text level the aim of the interpretative process is to work out encoded higher-order meanings, the challenge facing it is how the task can be expeditiously carried out, in view of the following constraints:

(1) The higher-level meaning is expressed by a (lower-level) structure already encoded with a lower-level meaning.
(2) The lower-level meaning cannot not be ruled out (as in cases of ambiguity), but has to be retained and integrated to the higher-level meaning, to which it stands in a hierarchical relation. Which aspect of the lower-level meaning is to be carried over to the higher level is a further problem, exhibited by argumentative acts.
(3) The identification of the higher-level meaning also requires quick access to relevant contextual cues.
(4) In the case of argumentational meaning, the identification of the higher-level meaning involves that of the upcoming text.

Irrespective of whether a stimulus-response connection as such is achievable in any given case, if the adaptive value of increased efficiency makes it desirable for processing systems to develop more direct connections, the basic requirement they must satisfy is one of finding neural pathways susceptible of acting as shortcuts between input and output points at various stages. In time these pathways gradually strengthen through use and become established.

In the cases under study, how the interpretative process hits upon the shortcut between the lower-level structure and the higher-level meaning requires – in my view – the intervention of divergent thinking. As the capacity to inhibit the tendency to act on immediate correlative stimulus relationships, thereby redirecting attention to the sampling of alternative (usually higher-order) associations, divergent thinking would appear to offer a perfect fit to the situation just described. Indeed, it caters for constraint (1) by inhibiting attention from remaining on the direct association between lower-level structures and lower-level meanings, and shifting it to higher-level meanings (i.e. meanings pertaining to a domain of reference further away from the external world than the one associated with the lower level). Divergent thinking also caters for the constraints in (2). Insofar as it entails a change of perspective involving the use of egocentric information as a frame of reference, while shifting attention to higher-order hierarchical associations (Deacon 1997, pp.263–5), divergent thinking also makes it possible for lower-level meanings to be retained in memory (within the scope of attention, but outside the focus) and to become embedded in the higher-level meanings (to which they contribute intentional contents). In the case of argumentative acts, divergent thinking is hypothesized to have intervened twice. Thus, prior to the attentional shift to the higher level, which is characteristic of both illocutionary acts (in the more traditional sense) and argumentative acts, divergent thinking would also be responsible for the perceptual switch whereby propositional meaning, normally envisaged under a representational aspect, can be viewed in a categorizing capacity (argumentative acts, as construed here, involves a categorization of the situation described by the propositional meaning in terms of a complex category). Inasmuch as constraint (3) requires a widening of the scope of attention, thereby enabling peripheral – in the event, contextual – entities to be encompassed within it, I argued that divergent thinking would also be involved, since it implies a widening of the scope of attention: indeed the allocentric associations to which attention is shifted need not belong to a higher level of analysis.

That divergent thinking should be called upon time and time again is only to be expected:

> As a species we have become predisposed to use this one cognitive tool whenever an opportunity presents itself, because an inordinate amount of other processes in the brain has become vested in the prefrontal cortex (Deacon 1997, p.265).

As a pre-existing means, divergent thinking (cf. Chapter 1) has proved its adaptive value in subserving tasks requiring:

(a) the inhibition of the tendency to respond to present stimuli (e.g. by satisfying a desire for immediate gratification) to allow the pursuit of a more distant goal;
(b) a hierarchization or temporal ordering of goal-oriented behaviour sequences.

Furthermore, as Deacon convincingly argues, the first symbolic recoding, that marks the emergence of language as a symbolic system, would not have been possible without it.

Once this interpretative shortcut has been made available by divergent thinking, the next stage for it is to become established. This comes about through repeated use impacting on the neuronal groups that make up the pathway thus created. In virtue of this property they have of responding to excitation by altering synaptic strengths, these neurons gradually strengthen the pathway at hand. What it means for a shortcut to become established is that the activation pattern that informs it becomes encoded in memory, thereby facilitating access to the appropriate endpoint upon identification of the stimulus.

This takes us back to the third constraint the interpretative process is subject to, which concerns gaining quick access to contextual cues, in the event, SN conditions. As will be recalled, contextual conditions cannot be expected to exhibit a high degree of invariance, in the sense of occurring across situations as a full set of identical conditions. For this reason no shortcut can be envisaged, unless there is a reliable source for contextual conditions. Thus posed the problem bears a clear resemblance to the one that gave rise to the genetic assimilation of environmental constraints. Briefly, where survival depends on a quick and an appropriate response to naturally occurring conditions, inasmuch as the latter cannot be expected to present a sufficient degree of invariance to ensure that they can be reliably identified, this has to be remedied. Natural selection's way of ensuring a quick and appropriate response is to internalize environmental constraints, so that perceptual hypotheses about stimuli can be formed on the basis of scant and incomplete evidence, and appropriate responses triggered, once a direct connection is in place between perceptual and motor representations.

So, assuming that the problem facing the interpretative process in respect of SN conditions is also one of obtaining an invariant set of representations on the basis of highly variable contextual stimuli, the means called upon should also be the same. This means lies in an averaging mechanism inherent in the way stimuli are processed, which gives rise to prototypes. In other words, invariance does not so much derive from external conditions, as it is created by the processing system.

In addition to this shared problem, the case at hand has one of its own, linked to the existence of mental SN conditions. How do conditions that lack sensory attributes acquire invariance?

On the basis of available preexisting means, a first scenario was considered, which involves:

(1) sensory correlates for mental conditions occurring at the same time as other contextual stimuli;
(2) a preprocessing system which categorizes stimuli on the basis of past usefulness, thereby enabling relevant contextual cues to become differentiated from accidental ones;
(3) a priming mechanism, whereby prior exposure to a set of stimuli will bias a system to respond to subsequent exposures to any one of these stimuli by activating all the connected representations;
(4) a depth of processing effect, whereby conditions that are objects of attention make their way – from short-term memory, where they are represented – to associative memory, through the mediation of reciprocal connections;
(5) reciprocal connections between lower and higher-order systems, which makes it possible for preprocessing systems to receive feedback from higher-order systems and short-term and long-term memories to swap information.

This scenario, however, runs into difficulty, as collective sensory correlates cannot lead to invariant individual conditions. To remedy this situation, I proposed that invariance could still be the result of the averaging process, but operating, this time, on concepts these sensory correlates are associated with. This revised scenario requires an appropriate theory of concept formation, one available, as we shall see presently, from Edelman's T.N.G.S.

The fourth constraint is specific to argumentational meaning. The problem facing interpretative process in this case is how to go about identifying the upcoming text. Inasmuch as this situation is construable as a case of decision making, and that decision making at the linguistic level arguably has an impact on survival or the quality of this survival, I hypothesized that the case at hand may be rooted in the neurobiology and subserved by some (at least) of the means involved in semi-conscious and conscious models of decision making, including those evolved to ensure processing efficiency. Rephrased in terms of parameters arising from Damasio's model, the above problem becomes one of finding the means that would speed up the decision-making process required by the following situation: a linguistic representation of a state of affairs is made with a communicative intent to the effect that the interlocutor should, on the basis of this linguistic event, identify what would constitute an appropriate continuation for it. These means involve in the case of Damasio's model:

- an appropriate source of knowledge providing the know-how that facilitates the selection of response options;
- supporting structures and mechanisms underlying the acquisition, encoding in memory and display of that knowledge during decision making;
- devices evolved to speed up decision making at various stages (where 'devices' is taken to be new uses of preexisting structures and mechanisms).

In respect of the source of knowledge, I proposed that here we are dealing with an extension of Damasio's repository of factual knowledge, namely, background knowledge. This knowledge would indeed provide the enabling capacities that underlie the activation of individual skills based on factual knowledge. Where factual knowledge is acquired on the basis of personal experience, back-ground knowledge is acquired on the basis of collective experience, in the sense of the same types of experience occurring to individuals sharing the same physical and social environments, as well as the same biological make-up. As a further source decision making can draw upon, background knowledge would, similarly to factual knowledge, consist of complex categories, that is, of categories of situations calling for a decision paired with categories of appropriate response options. However, the two types of complex category differ in one important respect: while appropriate responses in the case of individual skills are contingently associated with initial situations, those corresponding to background knowledge would appear to have resulted from a 'dissociation' process, rather, (at least, in the first instance); if, as Searle claims, our knowledge of how to handle an object is not separate from our knowledge of what that object is, then a perception or a representation of that object necessarily evokes practices associated with it.

Turning to supporting structures and mechanisms, all knowledge, whether innate or acquired, appears to consist of dispositional representations, or potential patterns of neuron activity within groups of neurons. Dispositional representations on which innate knowledge is based are located in the hypothalamus, brain stem and limbic system, and control processes required for survival; those corresponding to acquired knowledge, including, presumably, background knowledge, would be distributed throughout the higher-order association cortices (further details will have to await the forthcoming presentation of Edelman's theory).

During decision making the mechanisms of basic attention and working memory are called upon to display the categorized knowledge under consideration. The former serves to maintain a given mental image in consciousness, the latter, to hold and separate images for a period of hundreds to thousands of milliseconds, thereby enabling the supporting representation to be reiterated.

The main device evolved to boost decision making at the social level would appear to be the somatic marker device, whereby negative or positive somatic states, arising from experience, come to mark response options as deleterious or advantageous (to varying degrees). This device acts as a pre-selecting process, which narrows down the range of options to be considered by the reasoning process. As will be recalled, this basic device gives rise to an alternative mechanism that bypasses somatic states. In this 'as if' mechanism, the prefrontal cortices and the amygdala signal directly to the somatosensory cortex to organize itself in the appropriate activity pattern. Inasmuch as these two mechanisms can operate covertly, as well as overtly, this exemplifies a further level of automatization.

The closest thing to somatic markers that comes to mind is this negative or positive attitudinal component found in certain lexical items (e.g. 'prudent' vs. 'cowardly', 'courageous' vs. 'foolhardy'), which, though associated with the initial situation, rather than the response options, does bias the choice of response made by the interpretative process. As suggested in Chapter 3, the parallel with the somatic marker device is even closer, if the above lexical items are taken to result from a decision reached by S on the basis of a typical type of initial situation. The emergence of the type of argumentational meaning under study does not appear – in any obvious way – to involve a somatic device, at least not in its basic version. However, one cannot rule out the possibility of an 'as if' mechanism: insofar as background knowledge was also acquired, somatic states would have been present at the acquisition stage. As frequently encountered situations became categorized, the need for such states would, accordingly, have decreased. By the time the corresponding categorized knowledge was fully in place to provide a basis for skills, an 'as if' mechanism would have been operating. Inasmuch as such a mechanism underpins non-linguistic decision making, it would be prior to the emergence of argumentational meaning, which, among other things, requires that the initial situation should be (linguistically) represented. Such representation would be an extreme case of what Damasio (1994, p.184) refers to as 'symbols' of somatic states on which, early on, decision-making strategies began depending.

Processing efficiency in the case of argumentational meaning is also subserved by a novel use of the mechanism of basic attention as exemplified by the fact that we are dealing with symbolic representations of situations. In other words, where in 'normal' decision making the usual display of categorized situations 'freely' draws upon the subject's store of categorized knowledge, in the case of 'linguistic' decision making, a constraint is placed on what categorized knowledge is to be accessed: indeed, the symbolic representation provided by S of a given situation operates as a further level of automatization, in constraining H's categorization of the initial situation.

The last issue I wish to raise concerns evaluation. In attempting to provide an account of the emergence of the meanings under study, I availed myself of known pre-existing means. However, which among these were actually co-opted in each particular case, and at which stage, are not questions likely to be easily resolved through testing, if only because the same outputs can be produced by different inputs to any given subsystem. In view of this situation, the alternative is to try and show that this account is consistent with a global theory of brain function, one which, in addition would be:

(a) compatible with the co-evolutionary assumptions presented in the introduction;
(b) broad ranging enough to connect the various levels of analysis and processes featured in the account, so as to provide a coherent picture of their interactions.

Furthermore it is desirable that such a theory should also shed some light on a key issue our account left out, namely, how concept formation arises from lower-level processes.

Edelman's theory of neuronal group selection

A theory that appears to meet the above requirements is Edelman's biological theory of consciousness, an extension of his theory of neuronal group selection (TNGS) (Edelman 1987, 1988). The TNGS seeks to:

> explain in neural terms the processes of perceptual categorization, generalization and memory, and [to show] how their interactions can mediate the continually changing relations between experience and novelty that lead to learning (Edelman 1989, p.43).

Its extension (TNGSx) is an account of 'perceptual experience and awareness as they serve to guide action'. What the latter sets out to do is to 'show what interactions lead from perceptual *categorization* to perceptual *experience* – that activity allowing value-category matching in terms of time or succession which eventually leads to consciousness' (Edelman 1989, p.95).

Before proceeding to a presentation of the TNGSx, we must touch on the link between language and consciousness. Edelman takes language to be 'a major clue to the emergence of higher-order consciousness – the capacity that sustains direct awareness related to plans' (Edelman 1989, p. 173). While further details will have to await an overall presentation of the TNGSx, both language (as a symbolic system) and higher-order consciousness (as a special form of memory) appear to have this in common: they free the animal from the constraints and distraction of its immediate experience (which is response driven), thereby allowing the development, in the former case, and reinforcement, in the latter, of an ability to construct a coherent internal model of present, past and future; such a model functions to counteract the influence of its immediate experience. The adaptive value of such an ability can be best appreciated in those instances where predator recognition is dependent on a series of cues scattered over a certain period of time. It also subserves the coordination of sequences of actions, as required in carrying out a goal. On the basis of what we saw in the previous chapter, language at the text level does more than simply provide symbolic representations of such coherent pictures: the various constraints on context construction and sequencing act as relational categories participating in an identification task, which is interpretation.

The TNGS was prompted by two observations:

(1) The world is unlabeled until it becomes categorized by organisms endowed with internal adaptive criteria.
(2) Even within the same species the nervous systems of animals (especially vertebrates) present a significant amount of individual variations.

These observations (reinforced by similar ones in evolution and immunology) led to 'neural Darwinism', or the application of population thinking to nervous systems. In other words, in adapting to an uncategorized world, the brain would act 'by a process of selection upon variance' (Edelman 1989, pp.40–1).

In viewing the brain as a selective system, the TNGS constitutes a rejection of information processing models of it. Such models do not consider the level of variability of both the world and nervous systems to be significant enough to rule out a functionalist model of their interactions (that is, one defined in terms of the internal functional organization, or pattern of cause-and-effect relations, rather than in terms of physical features). Such models take a physical input and process it through their sequence of causal relations to produce a physical output. According to the TNGS, the world is not a source of stimuli providing information to the brain through a communications channel, nor is the brain an information processor carrying out pre-established procedures.

To explain the brain's ability for perceptual categorization, generalization and memory, the TNGS postulates three basic mechanisms: developmental selection, experiential selection and reentrant mapping (Edelman 1989, p. 45–72).

The first process occurs at the level of anatomical variance to yield variant anatomical networks (or 'primary repertoires') in each individual. The surviving neuronal groups then undergo experiential selection, in the sense that, as a result of behaviour, certain populations of synapses are strengthened or weakened to give rise to 'secondary repertoires'. Reentrant mapping involves sheets of neurons (or 'maps'), corresponding to particular regions of the nervous system. These maps are related to sensory receptor sheets (such as the retina of the eye) and to other maps by means of 'reentry', a process of parallel signalling using connections that are generally reciprocal. Repeated encounters with a stimulus lead to patterns of neurons being selected in a mapped area. Following this selection subsequent activation of each neuronal group by the same type of stimulus will have the effect of activating all other groups to which it is reentrantly connected. A major strength of reentry is that, unlike a strictly hierarchical model, it can account for how the responses of functionally segregated maps operating in parallel can be coordinated, thereby obviating the need for a pre-established programme and a homunculus to guide it.

Reentry, which also links otherwise disjunctive maps, can give rise to such diverse functions as 'cross-modal construction', 'resolution of conflicts in various areas' and 'recursive synthesis'. The pairing of the function of 'occlusion of objects' and of that of 'motion in vision' is an example of cross-modal construction; conflict resolution intervenes when competing sensory signals are received; as for recursive synthesis, it operates, through repeated reentrant signalling, to allow – among other things – outputs from higher-order maps to be recycled to lower-order maps for use in the generation of further neuronal responses. Edelman insists that feedback is by no means the sole function of reentry. Reentry, crucially, also has a constructive function, inasmuch as the parallel sampling involved can range over spatially extended areas, and concerns selective systems in which information presents statistical variations in parallel channels.

As a result of past and ongoing reentrant activity among related neuronal groups, attendant connections between maps strengthen, giving rise to 'classification couples' based on the pairing of selected neuronal groups in two different maps. Reentry in a classification couple links patterns of responses across maps, thereby making it possible for responses to disjunctive characteristics in a stimulus to become associated in the responses of a higher-order map. This accounts for how certain particular features of an object can be connected with more general ones. By also providing a linkage to previous responses to combinations of local features, or feature correlations, reentry makes generalization[1] possible: how it achieves this is by bringing to bear on the categorization of an object not encountered before past responses to similar objects.

Classification couples provide the basis for global mapping, a distributed system involving:

> *multiple* reentrant local maps (both motor and sensory) that interact with non-mapped regions such as those of the brain stem, basal ganglia, hippocampus and parts of the cerebellum. The activity of global mapping connects neuronal groups selected in one set of local maps (as a result of the activity of feature selectors) to neuronal groups selected in other sets of maps (as a result of the correlation of features that is established, for example, by continuity of motion) (Edelman 1989, p.54).

As a key concept in the TNGS, global mapping based on reentry derives its significance from the fact that the corresponding process provides an explanation for how categorization is related to memory, both of which are necessary to learning (or 'acquired context dependent behavioral change' (Edelman 1989, p.93)). The basis for memory is provided by short- and long-term changes in synaptic strengths arising from previous reentrant activity across different maps. Although changes in synaptic efficacy provide a basis for memory, the latter is distinct from the former. 'Memory' in the TNGS refers to the 'specific enhancement of a previously established ability to categorize', which emerges from the continual changes in synaptic populations and global mappings. To the extent that perceptual categorization is subject to alteration through changes undergone by the structure and dynamics of the original neural population (under the influence of changing contexts), this leads to an essentially non-

1 Edelman (1989, pp.49–50) defines perception as 'the adaptive discrimination of an object or event from background or other objects or events.' Generalization, he construes as 'the treatment within any given context of a more or less diverse collection of such entities as equivalent'. Reentrant maps derive their generalizing capacities from a property of neuronal groups in repertoires, known as 'degeneracy'. It is in virtue of this property that different combinations of neuronal groups can yield the same or a similar categorical output, and a given group can participate in more than one signalling function.

replicative conception of memory (Edelman 1989, pp.109–10). Thus, while recall of a previously given response in its original form cannot be ruled out, a similar response to the original is deemed to be the norm. In light of continual alteration in global mappings, one would have expected responses of a different nature. That this should not be the case is to be put down to the 'degeneracy' of neuronal groups, this property whereby the same or similar outputs can be achieved through a variety of pathways. For that reason, memory in the TNGS is more accurately described as 'recategorical' than merely 'nonreplicative'. Inasmuch as many different neural systems are amenable to synaptic change, in basing the notion of memory on it, the TNGS is subscribing to the possibility that there are different kinds of memory, in the event, a memory underlying primary consciousness and devoted to value-category successions, a special memory system for the development of concepts, as well as interactive memory systems subserving language, among others.

Our next concern is with how perceptual categorization and recategorical memory are related to learning. Learning, which 'improves the organism's chances of satisfying homeostatic, appetive and consumatory needs' (Edelman 1989, p.57), is made possible by neural linkages between global mappings that subserve perceptual categoryization and memory, and the activity of hedonic centres and the limbic system. Through the operation of these linkages, behavioural responses with adaptive value can be brought to bear on current categorization.

Inherent in the above is the distinction between two parts of the nervous system, one dedicated to the categorization of signals arising (mainly) from the external world, and one whose raison d'être is to ensure the 'persistent dominance of adaptive homeostasis' (Edelman 1989, p.94). These two parts of the brain provides the basis for the ability to discriminate between 'self'[2] and 'nonself', a necessary condition to the emergence of primary consciousness, the other three

2 As used in the TNGS, 'self' is to be taken in a strictly biological sense and is 'fundamentally determined by the signaling activity of areas mediating homeostatic – autonomic, hedonic, neuronendocrine – brains functions' that are automatic and sustain survival (Edelman 1989, p.98).

being perceptual categorization, memory, and learning. In addition to these four necessary conditions, two sufficient conditions are required. These are:

(a) special memory repertoires dedicated to the storage of past value-category associations;
(b) reentrant pathways allowing signalling between these special memory repertoires and those involved in current perceptual categorizations.

The value-dominated memory is based on a conceptual system capable of relational categorization (Edelman 1989, pp.140–6). The relations concerned are extremely varied in that they may involve those based on immediate perceptual categorizations and past categorizations, movement categories in respect to spatial and temporal references, values and categorizations of sensory experiences. Concepts differ from perceptual categories in being more abstract. Concept formation is, indeed, an ability to handle the relations between perceptual categories in the absence of appropriate perceptual stimuli. Furthermore, these categories may be based on past events. This high level of abstractedness is best exemplified in 'ontological concepts' corresponding to objects,[3] motions, hedonic states, classes. It is also manifest in concepts[4] schematizing general physical situations such as relations of containment, blockage, balance, superimposition, etc. Irrespective of whether it is capable of speech, an animal that possesses concepts is one whose actions are guided by an ability to categorize and generalize on relations, decide on sameness or difference and assess intentions on the basis of behaviour, and

3 The ability to recognize what an object is apparently depends on the possession of such concepts as those of 'cohesion, boundaries, substance and spatio-temporal continuity', as arising from perceptual experiences or experiences of moving objects (Edelman 1989, p.144). This view is based on E. Spelke's study ('Origins of physical knowledge' in *Thought and Language*, L. Weiskrantz (ed.), (pp.168–4). (Oxford, 1988))

4 Edelman (1989, p.141) surmises that these concepts provide bases for 'relations such as image schemata such as containment, in-out, blockage, etc', as studied by Johnson (1987) in relation to propositional structure, and Lakoff (1987).

consequences of actions. The brain areas involved in concept formation are hypothesized to include the frontal, parietal and temporal cortex, and the basal ganglia. By being reentrantly connected, these structures have the ability to:

- to restimulate portions of past global mappings, according to modality (or submodality) or types of movement;
- to combine and compare them.

As a result of such activity, cortical areas give rise to classifications of global mappings, which provide a basis for action. What facilitates the restimulation of common features of global mappings is that the various consequences of action can be abstracted and stored as synaptic changes by the frontal cortex. This same internal activity would make it possible for the frontal cortex to subserve generalizations (from global mappings, which can be modal or amodal).[5]

In this account of value-dominated categories reentrant connections between the above cortical areas and the hippocampus and the basal ganglia also deserve a special mention. Indeed the hippocampus matches perceptually significant inputs with subcortical signals arising from centres mediating internal values and hedonic states (Edelman 1989, p.129). As for the basal ganglia, insofar as their role is to correlate sensory and motor responses, they are believed to integrate signals received from several areas in terms of hedonically modulated goals and motivations, in the process of initiating motor responses (Edelman 1989, pp.137–9).

The hippocampus and the basal ganglia, together with the cerebellum, constitute what Edelman calls 'organs of succession' (Edelman 1989, pp.119–39). Indeed, they interact with the cortex to achieve the necessary linkage between spatial patterning and temporal change, and in so doing ensure the continuity of individual perceptual categories (e.g. in correlating, via reentry, the various signals corresponding to the features of an object) and their succession in

5 'Similar outputs can be achieved from mappings involving different modalities (e.g. vision and touch)' (Edelman 1989, p.144).

time. In virtue of this function, they provide the substrates on which coordinated action and the continuity of consciousness itself depend.

The problem of successive ordering begins with the categorization of motion. More specifically, it is a matter of how to achieve smooth succession of motions in gestures[6] and of gestures in 'synergies' (or classes of related gestures). While the succession of the motor signals is provided by the cortex, the smoothness of that succession itself arises from the activity of the cerebellum, which appears to have an essential role in timing and reflex control of successive movements. Together with the cortex it also determines the ordering of outputs to muscles. Smooth movement patterns (or synergies) are deemed by the TNGS to be essential to the continuity of perceptual categorization, in that they constitute the principal means of feature correlation.

The problem of ordering next arises in connection with short-term memory, which involves sequences of categorizations (as opposed to individual categorizations) and over longer time periods than those required by individual movements. If prior perceptual categorizations can be assumed, the first condition for the emergence of short-term memory is that there should be a means to achieve serial linkage between these categorizations. This is catered for by the activity of classification couples. The second condition concerns the ordered linkage of higher-order classification couples and n-tuples, which linkages must also be stabilized over a certain time period. This function is carried out by the hippocampus. The latter, as mentioned earlier, also mediates the modulation of this ordering by hedonic stimuli.

If perceptual categorization is to lead to learning, sensory responses must be linked to motor responses. This linkage, as will be recalled, is the responsibility of the basal ganglia.

As can be expected, the problem of ordering also arises in respect of goal oriented behaviours, which extend over longer periods of time (Edelman 1989, pp.151–72). In this case the main structure involved

6 Edelman (1989, pp.120–1) defines a gesture as 'one of the degenerate sets of all those coordinated movements that produce a particular pattern'.

in regulating the ordering is the prefrontal cortex of the frontal lobe. What makes this structure particularly suited for this role is that, in addition to having strong reentrant connections with other cortical regions, it has a capability for anticipatory set, arising from a short-term working memory, and is believed to have a key role in generating motor plans and converting them, via the basal ganglia, into motor programs.

Inasmuch as the categorizations it is involved in coordinating are temporally separate, the prefrontal cortex requires more than an appropriate short-term memory (that keeps the predetermined goal in mind): it has to have an ability to suppress interference, one arising from an attentional process. (This is consistent with an assumption introduced earlier that this structure is essential to the implementation of divergent thinking.)

On the basis of the varied nature of its connectivities and its central function in regulating the succession of goal-oriented actions, it is hardly surprising that the prefrontal cortex should be regarded as one of the concept formation centres. Indeed it gives rise to classifications of global mappings through its ability to 'abstract the various consequences of actions and store them as synaptic changes capable of *restimulating* common features of global mappings characteristic of different individual experiences' (Edelman 1989, p.145). Furthermore, through connections with the basal ganglia and the limbic system, it would also have the capability of relating values and perceptual categorizations. In other words, it is one of the loci for the conceptual or value-category memory. This linkage between category and value is what gives rise to learning, insofar as the adaptive responses based on them lead to changes in behaviour.

Perceptual experience, or primary consciousness, of objects and events results from the reentrant interaction between a value-dominated memory and ongoing perceptual categorizations and their short-term succession. Thus, perceptual experience of an object arises when previous conceptual categorizations interact with the outputs of the reentrant pathways to and from mapped receiving areas, namely a correlated set of signals corresponding to the various features of the object (these signals, as will be recalled, would have been registered by separate maps).

Because the memory involved is value-dominated, it constitutes a source of internal criteria whereby salience can be assigned to some events rather than others. This salience, which is based on the animal's adaptive needs, frees it from the demands of ongoing categorization.

Although a primary consciousness presupposes a self-nonself distinction and a behaviour influenced by long-term learning, an animal endowed with it, and capable of perceiving and feeling still cannot make plans independent of its immediate experiences. Freedom from the present does not arise until the animal is capable of being' conscious of being conscious', which entails the ability to symbolize the self- nonself distinction, or higher-consciousness. Such an ability requires the emergence of concepts of self and nonself that can only arise through social interaction. 'An awareness of the actions of the self and of others in various temporal successions entails a linkage to past action. This, in turn, entails a symbolic representation of past-action via long-term memory' (Edelman 1989, p.187).

Possession of higher-order consciousness, or at least of the beginnings of it, does not entail the possession of a true language. (Chimpanzees would be a case in point, as they have yet to evolve language despite their apparent ability to use individual symbols with some degree of semantic reference.) Language requires more than those areas of the brain concerned with concepts and the existing cortical apparatus involved in the categorization of global mappings and presyntax (or the ability to link concepts in a certain order), it also requires specialized developments related to verbal expression and recognition, which would have subsequently evolved into what is now recognized as the Broca's and Wernicke's areas. True syntax emerges when there is a sufficient phonology for words and sentences to become symbols for concepts.

With the abilities they involve, higher-order consciousness and language allow the animal to construct a coherent image of past, present and future, which is essential to its capacity for planned behaviours that are goal-oriented.

The types of meaning we have been concerned with would have emerged much later: the same mechanism the TNGSx puts forward to account for earlier developments of both lexicon and grammar would also be responsible for them. This mechanism is one of conceptual categorization, whereby the conceptual apparatus works recursively to:

> classify the various productions of language themselves – morphemes, words, sentences – as entities to be categorized and recombined without any *necessary* further reference to their initial origins or to their basis in perception, learning and social transmission (Edelman 1989, p.174).

Before turning our attention to how the various proposals advanced in the previous chapters may fit into the overall picture provided by the TNGSx, what further adaptive advantage may derive from the possession of a fully developed language, as opposed to an incipient higher-order consciousness? The possession of both, Edelman links with the emergence of the concepts of past and future and the attendant ability to construct a unified internal model of past, present and future. Such an ability is deemed to be crucial to the planning of goal-oriented behaviours related to survival.

Meanings from the perspective of the TNGS

On the basis of what we know of meanings at the text level, the question should be rephrased in terms of the capacity to provide external representations (i.e. as based on a conventional symbolic system) of internal models. Such representations, on account of their level of stability, may be placed on the same continuum as long-term memory. However, a more significant advantage lies in the possibility of involving others in the elaboration of one's plans. In that respect, discourse is about modelling one's internal states for the benefit of others, or rather, with a view to eliciting from them certain actions connected with one's own survival. Furthermore, we have seen that the means involved consist, for the most part, in facilitating mental representations of the desirable internal model in the interlocutor. What in previous chapters was referred to as 'constructed context' can be subsumed, within TNGSx, under the past component of this unified picture of past, present and future. 'Past' stands for value-category matchings, which determine the level of salience of ongoing categorizations, and classification couples linking sensory categorizations and motor responses, which determine future outcomes.

The fact that linguistic means are involved in inducing in others various operations leading to a unified picture of S's intentions is, in itself, a source of a further difference between higher-order consciousness and language: external representations may be seen to be carrying out a function similar to that of the attentional process. In privileging certain internal states over others as objects of representation, they act as a selecting mechanism to prevent H's attention from shifting to simultaneous processes (as connected to his own present concerns and values), which may interfere with the achievement of S's goal.

As our last point, let us consider how our proposals fit in with the TNGSx. For our purposes, the key notions are those of synaptic efficacy (as a basis for storage in memory) global mapping between different modalities, classification couples, reentry, value-category or conceptual memory (in its capacity to influence perceptual categorization and to further categorize its own concepts) and the various mechanisms whereby the ordering of successive categorizations is achieved.

Insofar as global mapping is a cross-modality process, this lends support to our proposal that illocutionary acts may be an instance of illocutionary value becoming connected with the motor representation of the appropriate utterance.

Classification couples (with what they presuppose in terms of supporting mechanisms) provide a basis for both complex categories, which derive from background knowledge, and Damasio's categorized factual knowledge, which arises from individual experience.

Recursive synthesis, which is a function of reentry, furthers our understanding of how higher-order meanings can become encoded in lower-level structures; indeed it allows higher-order constructs to be used as inputs to lower-order maps. This nicely complements our answer in terms of divergent thinking (which is compatible with

Edelman's view of the adaptive value of attention), and established pathways.

The supporting mechanisms which provide the basis for the emergence of primary consciousness offer a more satisfactory account of SN conditions than the ones proposed in Chapter 2. Indeed, how mental conditions acquire invariance poses no particular problem within the TNGSx, which offers an appropriate theory of concept formation. Concepts, as construed by Edelman, are relational categories, which entails that their emergence is mediated (among other things) by the categorization process, from which they derive their invariance. This invariance would be subsequently passed down to mental SN conditions, when conceptual information about them is activated in the course of relevant tasks. Using the TNGS as a general framework would also provide a simpler scenario for the problem posed by the encoding of the above conditions. Indeed, they could, arguably, acquire their invariance directly, through internal (i.e. conceptual) inputs.

The account of argumentational meaning based on Damasio's model of decision making would not have much to gain from being recast in terms of the TNGSx, which, in many respects, it closely resembles. Nonetheless, construing Damasio's repository of categorized factual knowledge as resulting from the operations of global mappings and a value-category memory enables us to see yet another layer of connections between the various issues at hand.

Although our concern was never directly with concept formation, it forms an integral part of the emergence of the meanings under study. Edelman's proposals, therefore, go some way towards filling this gap, in the sense that the conceptual apparatus he envisages can work recursively to create new concepts on the basis of existing ones. In view of their domains of reference, the types of concepts we have been dealing with – unlike those representing basic things, events, and relationships – are not likely to have arisen prior to the emergence of language.

Bibliography

Allport, A. 1987. 'Selection for action: Some behavioral and neurophysiological considerations for attention and action.' In H. Heuer and A.F. Sanders (eds.) *Perspectives on Perception and Action*. Hillsdale, N.J.: Lawrence Elbaum Associates, 395–419.

Anscombre, J.C. 1979. 'Délocutivité benvénistienne, délocutivité généralisée et performativité'. *Langue Française 42*, 69–84.

——(ed.). 1995. *Théorie des Topoï*. Paris: Kimé.

——1995. 'Topique or not topique: formes topiques intrinsèques et formes topiques extrinsèques'. *Journal of Pragmatics 24*, 115–41.

Austin, J.L. 1962. *How to do things with words*. Oxford: Clarendon Press.

Blakemore, C. and Greenfield, S. (eds.) 1987. *Mindwaves: Thoughts on Intelligence, Identity and Consciousness*. New York: Blackwells.

Blakemore, D. 1987. *Semantic Constraints on Relevance*. Oxford: Basil Blackwell.

Benveniste, E. 1966. *Problèmes de Linguistique Générale*. Paris: Gallimard.

Bruxelles, S., Raccah, P.Y., and Ducrot, O. 1993. 'Argumentation et champs topiques lexicaux'. *Cahiers de Praxématique 21*, 88–104.

Bruxelles, S., Ducrot, O., and Raccah, P.Y. 1995. 'Argumentation and the lexical topical fields'. *Journal of Pragmatics 24*, 97–114.

Chomsky, N. 1966. *Cartesian Linguistics*. New York: Harper and Row.

——1972. *Language and Mind*. (Extended ed.). New York: Harcourt Brace World.

——1986. *Knowledge and Language: Its Nature, Origin, and Use*. New York: Praeger.

Cohen, N.J. and Squire, L.R. 1980. 'Preserved learning and retention of pattern-analyzing skill in amnesia: Dissociation of knowing how and knowing that'. *Science 210*, 207–9.

Croft, W. 2000. 'Lexical and grammatical meaning'. In G. Booij, C. Lehman, J. Mugdan in collaboration with K. Kosselheim and S. Skopteas. *An International Handbook on Inflection and Word-Formation. Vol. 1*. Berlin, New York: Walter de Gruyter, 257–63.

Damasio, A. 1994. *Descartes' Error*. New York: The Free Press.

Deacon, W.T. 1997. *The Symbolic Species*. New York: Norton.

Dennett, D.C. 1970. *Brainstorms*. Montgomery, VT: Bradford Books.

——1991. *Conciousness Explained.* Boston, Toronto, London: Little, Brown, and Company.

Ducrot, O. 1972. *Dire et ne pas dire*. Paris: Editions de Minuit.

——1975. 'Je trouve que…' *Semantikos 1,1*, 63–88.

——1980. 'Analyses pragmatiques'. *Communications 32*, 11–60.

——1984. *Le Dire et le Dit*. Paris: Editions de Minuit.

——1995. 'Topoï et formes topiques'. In J.C. Anscombre (ed.). *Théorie des Topoï*. Paris: Kimé, 85–99

Eccles, J.C. (ed). 1966. *Brain and Conscious Experience*. New York: Springer-Verlag.

Edelman, D.G. 1987. *Neural Darwinism: The Theory of Neuronal Group Selection*. New York: Basic Books.

——1989. *The Remembered Present*. New York: Basic Books.

Elman, J.L., Bates, E.A., Johnson, M.H., Karmiloff-Smith, A., Parisi, D. and Plunkett, K. (eds.), 1996. *Rethinking Innateness.* Cambridge MA: The MIT Press.

Fischer, K. *Discourse particles*. To appear.

Fodor, J.A. 1983. *The Modularity of Mind.* Cambridge MA: The MIT Press.

——*Psychosemantics: The problem of Meaning and the Philosophy of Mind.* Cambridge MA: The MIT Press.

Fuster, J. 1980. *The Prefrontal Cortex*. New York: Raven Press.

Givón, T. 1979. *Understanding Grammar*. New York: Academic Press.

Goldman-Rakic, P. 1987. 'Circuitry of primate prefrontal center and regulation of behavior by representational memory'. In F. Plum (Vol. Ed.) and V.B. Mountcastle (Sec. Ed.). *Handbook of physiology: The nervous system, Vol. 5: Higher functions of the brain*. Bethesda, Md: American Physiological Society, 353–417.

Guilford, J. 1967. *The Nature of Human Intelligence*. New York: McGraw-Hill.

Harris, M. 1978. *The Evolution of French Syntax*. New York: Longman.

Heine, B., Claudi, U., Hünemeyer, F. 1991. *Grammaticalization*. Chicago: University of Chicago Press.

Hopper, P. and Traugott, E. Closs. 1993. *Grammaticalization*. Cambridge: Cambridge University Press.

Johnson, M. 1987. *The Body in the Mind: The Bodily Basis of Meaning, Imagination and Reason*. Chicago: Chicago Unversity Press.

Johnson-Laird, P.N. 1983. *Mental Models*. Cambridge MA: Harvard University Press.

Kandel, E.R., Schwartz, J.H. and Jessell, T. M. (eds). 2000. *Principles of Neural Science*. Fourth Editions. McGraw-Hill.

Kirby, Simon. 1999. *Function, Selection and Innateness*. Oxford: Oxford University Press.

Kosslyn, S.M. and Koenig, O. 1995. *Wet Mind*. New York: The Free Press.

Lamb, S.M. 1998. *Pathways of the Brain*. Amsterdam: John Benjamins.

Langacker, R.W. 1990. 'Subjectification'. Cognitive *Linguistics 1*, 5–38.

——1991. *Foundations of Cognitive Linguistics*, Vol.2. Stanford: Standford University press.

Lakoff, G. 1987. *Women, Fire and Dangerous Things: What Categories Reveal about the Mind.* Chicago: University of Chicago Press.

Lakoff, R. 1971. 'If's, and's, and but's about conjunction'. In J.C. Fillmore and D.T. Langandoen (eds.). *Studies in Linguistic Semantics*. New York: Holt, Rinehart and Winston, 115–50.

Lashley, K.S. 1951. 'The Problem of serial order in behavior.' In L.A. Jeffress (ed.). *Cerebral Mechanisms in Behavior*. New York: Wiley, 325–34.

Maclaury, R.E. 1997. *Color and Cognition in mesoamerica: Constructing Categories as Vantages*. Austin: University of Texas Press.

Marmaridou, S.S.A. 2000. *Pragmatic Meaning and Cognition*. Amsterdam, Philadephia: John Benjamins.

Mishkin, M. 1978. 'Memory in monkeys severely impaired by combined but not separate removal of amygdala and hippocampus', *Nature 273*, 297–8.

Mishkin, M. and Appenzeller, T. 1987. 'The anatomy of memory'. *Scientific American 256*, 62–71.

Mishkin, M., Malamut, B. and Bachevalier, J. 1984. 'Memories and habits: Two neural systems'. In G. Lynch, J.L. McGaugh and N.M. Weinberger (eds). *Neurobiology of learning and memory*. New York: Guilford Press, 65–77.

Nagel, T. 1974. 'What is it like to be a bat?' *Philos. Rev. 1974*, 435–50.

——1986. *The View from Nowhere*. New York: Cambridge University Press.

Neumann, O. 1987. 'Beyond capacity: A functional view of attention.' In H. Heuer and A.F. Sanders (eds.). *Perspectives on Perception and Action*. Hillsdale, N.J.: Lawrence Erlbaum Associates, 361–94.

Nyan, T. 1993. 'Au moins'. Paper presented at a worshop on Pragmatics, Discourse and 'Enonciation', at the University of Salford (November 1993).

——1998. *Metalinguistic Operators, with reference to French*. Bern: Peter Lang.

——1999a. 'Simplement.' *Argumentation 13*, 275–95

——1999b. 'Vers un schéma de la différence: le cas de mais.' *Journal of French Language Studies 9*, 211–38.

——2002. 'Vantage Theory and Innateness'. *Language Sciences* 24 : 705–12

——2003. 'From procedural meaning to processing efficiency'. In K. Fisher (ed.) *Discourse Particles*. To appear.

OED: Oxford English Dictionary, 2nd edition.

Perrett, D.I., Smith, P.A., Potter, D.D., Mistlin, A.J., Head, A.S., Milner, A.D., and Jeeves, M.A. 1985. 'Visual cells in temporal cortex sensitive to face view and gaze direction'. *Proceedings of the Royal Society of London, Series B223*, 293–317.

Pinker, S. and Prince, A. 1988. 'On language and connectivism. Analysis of a parallel distributed processing model of language acquisition'. *Cognition 28*, 73–193.

Popper, K.R., and Eccles, J.C. 1981. *The Self and its Brain*. Berlin: Springer-Verlag, 28–48.

Ramat, A. and Hopper, P. 1998. *The Limits of Grammaticalization*. Amsterdam: John Benjamins.

Récanati, F. 1987. *Les énoncés performatifs*. Paris: Editions de Minuit.

Rosch, E. 1978. 'Principles of Categorization'. In E. Rosch, B. Lloyd (eds.). *Cognition and Categorization*. Hillsdale, N.J.: Laurence Erlbaum Associates, 28–48.

Schacter, D.L., Cooper, L.A. and Delaney, S.M. 1990. 'Implicit memory for unfamiliar objects depends on access to structural description'. *Journal of Experimental Psychology: General, 119*, 5–26.

Schacter, D.L. and Graf, P. 1986. 'Effects of elaborative processing on implicit and explicit memory for new associations'. *Journal of Experimental Psychology: Learning, Memory and Cognition 12*, 432–44.

Searle, J.R. 1969. *Speech Acts*. Cambridge MA: Cambridge University Press.

——1979. *Expressions and Meaning*. Cambridge University Press.

——1980. 'Minds, Brains and Programs'. *Behavioral and Brain Sciences 3,* 417–24

——1983. *Intentionality*. Cambridge University Press.

——1989. 'How performatives work' *Linguistics and Philosophy 12*, 535–58.

——1992. *The Rediscovery of the Mind*. Cambridge MA: The MIT Press.

——1995. *The Construction of Social Reality*. New York: The Free Press.

——2001. *Rationality in Action*. Cambridge MA: The MIT Press.

Spelke, E. 1988. 'Origins of physical knowledge.' In L. Weiskrantz (ed.). *Thought and Language*. Oxford: Clarendon, 168–84.

Sweetser, E. 1986. *From Etymology to Pragmatics*. Cambridge University Press.

Taylor, J.R. 1989. *Linguistic Categorization*. Oxford University Press.

Traugott, E. 1999. 'The role of pragmatics in semantic change.' In J. Verschueren (ed.). *Pragmatics in 1998: Selected papers from the 6th International Pragmatics Conference, vol. II*. Antwerp: International Pragmatics Association, 93–102.

Traugott, E. Closs and Dasher, R. 1987. 'On the historical relations between mental and speech act verbs in English and Japanese'. In Ramat, A., Gracelene et al. (eds). *Papers from the 7th International Conference on Historical Linguistics*. Amsterdam: John Benjamins, 560–73.

Traugott, E. Closs and Heine, B. 1991. *Approaches to Grammaticalization, Vol. 1 & 2*. Amsterdam: John Benjamins.

TLS: *Trésor de la Language Française: dictionnaire de la langue du XIXe et du XXe siècle*. Paris: Editions du Centre National de la Recherche Scientifique.

Van Eckerdt, B. 1995. *What is Cognitive Science*? Cambridge MA: The MIT Press.

Van Essen, D.C. and Maunsell, J.H. 1983. 'Hierarchical organization and functional streams in the visual cortex'. *Trends in Neurosciences 6*, 370–75.

Index

Ability 18, 31, 67 n.5, 69, 92, 93, 105, 107, 139, 140, 142, 166, 167, 175, 176
Act of argumentation 94, 112–115, 117–119, 122, 123, 125, 128, 129, 131, 134, 140, 159
Act of categorization 112
Act of communication 95, 112, 114
Accuracy 83 – see also processing efficiency
Activation patterns (activity patterns, patterns of activation) 73, 75, 81, 84, 85, 101, 102, 110, 153, 160, 164, 167
Adaptation (language) 18, 29, 59, 89, 115
Adaptive advantage 175, 176
Adaptive needs 175
Adaptive value 156, 166, 170
Alternatives (higher-order) – see also divergent thinking
Amodal systems 172
Amygdala 99, 101, 164
Angle of viewing 42, 43, 128, 155
  subjective 23, 24, 54, 127
Anscombre 14, 89, 90, 91, 141
Anterior cingulate 101
Anticipation 72, 84
Appenzeller 36 n.6, 84
As if mechanism 101, 164
Aspect 22, 39, 55, 124, 125, 134
  categorizing 116, 125, 126, 128, 133, 151
  causal 23
  interrelational 24
  representational 44, 54, 151
Associations
  allocentric 19, 31, 64, 65, 121, 124, 151, 159
  egocentric 19, 31, 65, 121, 124, 151, 159
  higher-order 27, 149
  lower-order 25
  value-category 170
Attention
  basic 103, 163, 164
  and interference, 174, 177
  focus of 22, 30, 38, 41, 101, 103, 147, 151
  attentional mechanism 101
  attentional shift 27, 29, 30–32, 46, 55, 64, 70, 121, 123, 131–133, 136, 150, 151, 159
  attentional threshold 67
  attention window 128
Attitudinal component 103, 164
Attribute (perceptual, sensory) 72, 74, 78, 82
Austin 37
Automatization 20, 24, 69, 70, 83, 87, 97, 98, 103, 123, 124, 136, 157, 164
  levels of 103, 108, 112, 120, 165
Averaging process 73, 161, 162 – see also generalization
Awareness (of information) – see also short-term memory
AWL 89, 90–92, 112, 113, 139, 153, 155

Bachevalier 36 n.6, 84
Background 92, 93, 95, 105, 107, 142, 143
  cultural background 93, 143
  deep background 106, 109, 143

local background 92, 105, 109, 117, 147
Basal forebrain 36 n.6, 101
Basal ganglia 84, 169, 171, 172, 174
Behaviour
flexible, adaptive 65
goal-oriented 160, 173, 175
Behavioural change 169, 174
Belief 32, 33, 41, 44, 76, 105, 125, 143
Benveniste 52
Bias – see also learning
selectional 17
biasing job 101
biasing system 103
Blakemore 25
Body state 99
Brain
and language 15–20, 66, 67
as information processor 167
as a selective system 167
functional nature of 167
modular nature of 86
Brain areas
mapped 167, 169
nonmapped 169
Brain stem 19, 101, 110, 163, 169
Broca's area 175
Bruxelles 90, 103
Buffer
sentence structure 75, 81, 86
visual 75

Canalization (of adaptive behaviour) 61
Capacity
categorizing 115, 116, 120, 125, 127–129, 134, 143, 147, 151, 152, 159
inhibitive 68, 158
nonrepresentational 75 n.13, 105, 106
representational 54, 120, 122, 127, 128, 132, 141, 145, 151, 152, 153
to generalize 73, 84
Categorization 141, 142
conceptual 175
ongoing 175, 176
past 171
perceptual 86, 100, 166–169, 173, 174, 177
processes 82, 87, 115
relational 171, 178
of motion 173
of sensory experiences 171
Category 73
complex 94–96, 105, 107, 112, 116, 117, 119, 120, 128, 133, 134, 140, 141, 145, 149, 159, 177
movement 171
perceptual 80, 169, 170
relational 166, 178
of situations calling for a decision 94, 103
of response-options 94, 103, 119
Category judgment 115–117, 128, 130, 133
Centre stage 24 – see also subjectification
Cerebellum 169, 172, 173
Cerebral hemispheres 71, 88 – see also neural substrate and segregation
Chomsky 15, 16, 18,
Classification couples 169, 173, 176, 177
Coarse coding 84, 85
Coding 25 – see also recoding
lexical 21, 25
syntactic 25
modalities 21
of illocutionary value 53, 55, 58
modes of 46
object of 29
Co-evolution 15–20, 67
Co-evolutionary assumptions 15–20, 156
Cohen 84
Colour 84

Commitment 113, 114, 118, 129, 144, 144
to category judgement 116, 117, 126, 142
to direction of fit 35, 44, 48, 49, 117
Communication
among intimates 22
among strangers 22
communication needs 20–26
communication situation 21, 35, 94, 95, 108, 112, 119
modes of 18
pragmatic mode of 21
syntactic mode of 21
communicative contents 21–25
communicative context 23
communicative goal 2, 32, 60, 95
communicative intention 25, 107
communicative pressure 25, 157
Concept 32 n.2
ontological 171
concept formation 82, 165, 171, 174, 178
loci 171
and perceptual category 171
Conceptual structure 24, 112, 138
Conceptual system 171
Concurrent processing 71, 84 – see also parallel processing
Conditions
of satisfaction 24, 33, 39, 44, 47, 54, 55, 60, 62, 63, 76 n.13, 106, 117, 133, 134, 136, 137
essential 62
external 65, 72, 161
idealized 72
individual 80, 82
preparatory 13, 63
mental 72, 74–76, 77, 81, 82
self-referential causality 33, 79
sincerity 14, 33, 62, 63, 77, 123
SN conditions 13, 59–87, 141, 168, 178 – see also built-in context
Connections
reciprocal 80, 86, 161, 167
re-entrant 167
Consciousness 102, 103
higher-order 166
primary 170
Constraint
on continuation 13, 14, 24, 92, 94, 103, 112, 120, 130, 132, 146, 139
on sequencing 13
on response 24, 60, 138
argumentional 94, 137, 138
environmental 59, 61 – see also internalization
illocutionary, 137, 138
weak 103
arising from lexical items, 138
Constraint satisfaction 14, 85
Content
communicative 21–25
intentional 32, 33, 35, 39, 106, 122, 124, 125, 129 nonrepresentional 105 – see also intentionality
propositional 32, 37, 42, 68, 76, 96, 122
representative 32
Context
of relevance 25
of utterance 58, 62
adaptive 67
built-in context 60, 74, 79
constructed 134, 176
encoded context of evaluation 24
idealized 24, 60, 70
learning 67
linguistic 72
mental 60
pre-programmed 72
situational 51
Context free situation 90
Continuity
of consciousness 172
of individual perception 172

of motion 169
Convergence zones 97, 100, 110 – see also prefrontal cortex
Co-option 27, 29, 32, 62, 83, 87, 113 n.13, 156, 165
Correlate (sensory) 79, 80, 83, 161, 162
Cortex
higher-order association cortices 110, 163
frontal 171, 172
motor 84
parietal 171
prefrontal 18, 19, 64 n.3, 67, 69, 100, 160, 173, 174
sensory 110
somatosensory 100, 101, 164
temporal 171
Cortical regions 18, 19, 172, 173
Co-text 27, 112, 119, 126
Croft 31, 121
Cross-modal construction 168

Damasio 89, 94–105, 107, 110–112, 119, 154, 162, 164, 177, 178
Dasher 39, 40, 51
Data 155, 156
Deacon 15–20, 29, 31, 61, 64, 66, 67, 69–71, 88, 146, 159, 160
Decision-making 27, 112, 119, 165
induced 96, 120
models of 89, 94–98, 113, 162, 164
process of 81 n.16, 102, 107, 150, 157 – see also complex categories and somatic markers
Degeneracy 169, 170, 173 – see also generalization and averaging process
Delocutivity 52–56, 112
Depth of processing effect 81, 161
Derivational processes 15, 51
Desire 33, 105
Direction of fit 38, 44, 45, 47, 48, 76 n.13, 152
Directionality of shift 31, 34, 39
Discourse 14, 22, 25
Discourse markers 24, 25, 103
Divergent thinking
and delocutivity 53
and higher-level meanings 27, 87
and illocutionary acts 31–36, 46, 51, 113
and modes of reference 19
and symbolic recoding 29, 30
and argumentational meaning 94, 122, 132, 134, 146
tasks involving divergent thinking 19, 121
uses of 63, 64, 65, 68, 69, 136, 137, 154 – see also mnemonic strategy, prefrontal cortex
criteria 31–36, 120
Domain
of discourse 147
of reference 2, 34, 43, 44–46, 118, 121, 122
abstract 31 n.2, 32
conceptual 31, 121
intersubjective 117
metalinguistic 25
subjective 21–23, 118, 134
target 122, 128, 134
Ducrot 25, 51, 52, 56, 57, 90

Edelman 36 n.6, 82, 83, 136, 162, 163, 165, 168, 170, 173–178
Edges (processing of) 86
Emotions
primary 98–100
secondary 98–100
Encoding 43, 59, 60, 62, 63, 70, 78, 82, 93, 94, 12, 113 – see also recoding
of associations 136, 178 – see also stimulus-response connection
of motor representation 177
of prototypes – see averaging process and generalization
degrees of 14, 34, 129

explicit memory encoding 76, 81, 84
linguistic 13 n.1, 69 n.7
perceptual 36 n.6
pre-encoded sentence 72
Environmental constraints 27, 160, 161 – see also internalization
Evaluation 165
Evolution
Baldwinian 66 n.4
Darwinian 66 n.4
Exemplar 73

Feature correlation 173
Feedback 80 – see also reciprocal connection
Feeling 99
Fischer 25, 103 n.11
Fodor 86 n.19
Frontal lobe 81 n.16
Functionalist view (of the brain) 167 – see also homunculus, modularity
Fuster 19 n.3

Generality requirement 126, 127, 144
Genetic assimilation 61, 65, 66, 160 – see also internalization
Generalization process 167, 168, 172 – see also averaging process, degeneracy, perceptual subsystems
generalizing tendency 74
capacity to generalize 73, 84, 164, 167, 169
Gesture 173
Givón 21–25, 157
Global brain theory 165
Global mapping 169, 172, 174, 175, 177, 178
Grammar 16, 175

Hedonic centres 170
Hedonic states 172
Higher-order systems (vs. lower-order) 80, 86
Hippocampus 36 n.6, 169, 172, 173
Historical Linguistics 14, 39, 51
Homunculus 136, 167 – see also functionalist view
Hunger state 97
Hypothalamus 97, 101, 110, 163
Illocutionary act 2, 35, 52, 53, 116, 117, 122, 177
assertive 14, 26, 32–36, 3, 40, 49, 123, 127
commissive 14, 33, 36–39
declarative 34
directive 33, 38
expressives 46–55
request 30, 38
logical structure of 33, 42, 51
point of 41
with a declarative structure 36–58
Illocutionary force 32, 33, 37
Illocutionary value 29, 30, 43, 59, 177
Image (internal) – see also primary consciousness
auditory 100
mental 81, 100–103, 164
perceptual 110
visual 100
Image schemata 171
Implicature 13 n., 30
Information
processing 69
processing model of the brain 167 – see also functionalist view
contextual 27, 35, 85
metalinguistic 26
swapping of 81, 86
Innate biological preferences 100, 101
Innate regulatory dispositions 101
Innateness 15–18
Inner values – see also hedonic centres
Instantiation (degree of) 26
Intention 41, 63, 105
subsidiary 106
logical structure of

Intentional content 40, 44, 106, 132, 133, 150
Intentional object 40, 42
Intentional states 106, 136, 140, 142 – see also mental states
pre-intentional 136 – see also the Network, representation, background
Intentionality 32, 42, 60 n.2, 122
structure of 42, 123
Internalization 61, 62, 64, 65
Internal criteria 74
Internal values 172
Interpretive problem 26, 93, 94, 112, 122
Interpretive process 69 n.6, 120, 178
Invariance 64, 72–78, 80, 160 – see also generalization, prototypes, activation patterns
source of 68, 69, 73
invariant features 61, 72, 162

Johnson 171 n.4

Know-how 75 n. 13, 97, 105, 108–111, 119, 140, 143, 162
Know-that 75 n.13, 110, 111
Knowledge
acquired 110, 163
background 21, 24, 27, 75, 108–110, 127, 144, 151, 162–165
categorized 94, 95, 101, 108, 133, 163, 164, 178 – see also complex categories
contextual 77, 86 – see also implicit and explicit
explicit 77
factual 105–108, 112, 162, 163
implicit 75, 76, 80 – see also background and stimulus-response connection system, skills
innate 110, 163
linguistic 18
Knowledge source problem 93, 93

Koenig 36 n.6, 71, 73–77, 80, 81, 84, 85, 87, 136, 152
Kosslyn 36 n.6, 71, 73–77, 80, 81, 84, 85, 87, 136, 152

Lakoff, George 170 n.2
Lakoff, Robin 25
Langacker 32 n.2, 121 n.13
Language acquisition 16
Language adaptation 59
Language change 14–17, 66, 68
Language competence 15
Language evolution 17, 21–25 – see also higher-order consciousness
Language universals 17, 66, 67
Learning 99
bias 17, 66
predisposition 19
and adaptation 163, 170, 173
and contextual pressures 169
symbols 31
Levinson 13 n.1
Lexicon 175
Limbic system 99, 100, 110, 163, 170, 174
Linguistic convention 13 n., 34, 25, 65, 124
Lower-order systems (vs. higher-order) 165

Mais 91
Malamut 36 n.6, 84
Maps 167, 168, 177
local 169
sensory 167
global mapping 167, 169 – see also mapped, unmapped areas of the brain, re-entrant mapping
Markers
discourse 14 n.2
somatic 98, 99 – see also argumentational meaning
Matching process 73, 74

Maunsell 85
Meaning
argumentational 13–15, 93, 113, 120, 134, 140, 156, 162, 164
contextual 15
conventional 13 n.1
higher-level 26, 29, 30, 35, 36, 65, 89, 94, 126, 146, 150, 158, 177
illocutionary 35, 68, 86
interrelational 25
lexical 90
linguistic 42, 51
literal 106
lower-level 15, 26, 30, 36, 38, 86, 93, 141, 144, 147, 150, 158, 159
procedural 103 n.11, 124
propositional 14, 26, 27, 33–37, 44, 45, 52, 68, 113, 122, 124, 128, 157
sequenctial unit of 149, 150, 153
at the text level 13, 14, 9, 157, 166

Memory
amodal 77
associative 72, 76, 78, 80, 81, 161
conceptual 80, 170, 174, 177
explicit memory 87
implicit memory 140
explicit memory formation 78, 82
linguistic memories 170
long-term 27, 75, 80, 86, 161
modality specific (or unimodal) 77
motor 36 n.6
nonreplicative 169, 170
perceptual 170
prelinguistic 80
recategorical 169, 170
short-term 69, 75 n.12, 81, 161, 173, 174
unimodal 75
value-category 170, 174, 178
value-dominated 171
working 20, 81, 86, 102, 163 – see also sentence buffer

Membership criteria 125, 144, 145
Mental states 23, 33–35, 38, 41, 45, 78, 137 – see also psychological states
expression of 41
and illocutionary acts 40, 122
Methodological issues 155–165
Miskin 36 n.6
Mnemonic strategy 20, 87, 136 – see also divergent thinking
Modality 77, 172
Modularity (weak, strong) 86, 155
Monkey 61
Motion 85, 169, 173 – see also categorization and synergies
Motor plan 174
Motor program 174
Motor representation 36
Movement
pattern 173
successive 173
planning of 110

Natural selection 160
Needs
appetitive 170
consummatory 170
homeostatic 170
Nervous system 61, 166
Network 106
(the)Network 107, 109, 142, 143
recurrent network design 73
Neural Darwinism 166
Neural circuits 61, 101
Neural substrate 70, 71, 100
co-option of 71, 87, 88
Neuronal group selection – see TNGS
Neurons 163
connections between 36 n.6, 86
maps of – see also synaptic connections
Neuro-transmitter nuclei 101 n.7
Nonaccidental properties 73

Non-objectivist conception of lexical meaning 90
Nonself 170 – see also self
Nyan 25, 26

Obligation 38, 61, 83, 109, 114, 116, 117, 132, 142, 145
  of consistency 116
Observability 3, 36
Order
  successive ordering 172, 173
  organs of succession 172

Parallel processing 71, 84 – see also concurrent
Past, present, future (model of) 166, 176
Pathways 36, 136, 157–160 – see also shortcut, and synaptic efficacy
Pattern activation subsystem 73, 75, 84
Perceptual strategy 35
Perceptual subsystems 72, 77
Performativity 51, 54, 56
Perrett et al. 77 n.11
Perspective (change of) 3, 121
Point of view 42, 55
  shift in 22
Practices 92, 93, 103, 107, 109, 136, 140
Preconditions (of intentional states) 106, 142
Predispositions (genetic) 16. 17, 27, 61
Preferred continuations 14, 26, 89, 90, 105, 139
Pre-intentional assumption 105, 143
Preprocessing subsystem 80, 86, 161
Pre-selecting process 164
Presyntax 175
Priming 61, 74, 161
Processing efficiency 20, 34, 60, 67, 71, 86, 88, 136, 157, 162, 164
Processing requirement 70, 71, 103 n.11
Prototype 73, 161
Psychological mode 32
Psychological states 47, 48, 50, 51, 106 – see also mental states

Raccah 90, 103
Ranking criteria 102
Reasoning
  process 77, 98, 102, 164
  ability 87
Recanati 53 n.9
Recoding 20, 21, 29
Recurrent network 73
Recursive synthesis 168
Reentrant connections 172, 173
Reentry 167, 169, 171, 172, 177
Reference
  indexical 19
  symbolic 16, 17, 66
Reference point 24 45
Repertoire
  primary 167
  secondary 167
Representation 26, 136 – see also intentionality
  of the context 77
  auditory 71
  dispositional 97, 98, 100, 110, 111, 154, 163
  external 176, 177
  internal 176
  long-term memory 81
  mental 176
  motor 62, 70, 81, 143, 157
  perceptual 79, 106, 152, 157
  symbolic 20, 165, 166, 175
  visual 71
  and focus of attention 103
  forms of 106
  object of 106
Representational aspect – see aspect
Representational capacity 44 – see also categorizing capacity
Response 61
  adaptive 6, 17, 41

motor 172
canalization of adaptive 61
sensory 173
suppression of 69
Response-option 97–98, 107, 162, 164
Rosch 73

Searle 14, 32, 33, 37, 39, 40, 42, 47–49, 57, 58, 78, 92, 105, 106, 110, 113, 117, 118, 136, 139, 141, 143, 144, 163
Segregation principle 71 – see also neural substrate
Selection
Balwinian 66
Darwinian 66
developmental 167
experiential 167
natural 61, 66
process 17, 112
response 97, 98
Selectional pressure 17, 71
Self 170, 175 – see also nonself
Self-referentiality 58
Semantic layering 34, 128
Sentence structure buffer 81
Sentence level 71
Sequence of categorization 173
Sequencing 91
Serial linkage 172, 173
Sheet (sensory receptor) 167
Shortcut 20, 27, 34, 35, 64, 65, 89, 120, 126, 128, 134, 136, 154, 157 – see also pathways, and divergent thinking
Situation
categorized 165
schematized 25, 103
speech 25
of reference 113, 115, 116, 126, 128, 134
Skills 76 n.13, 77, 81, 82, 115, 107, 108, 111, 140, 164 – see also implicit knowledge, Background, stimulus-response connection system, pre-intentional assumptions
Somatic markers 99, 163
Somatic states 101, 164
Speech act 14, 23, 30, 32, 33, 40, 71, 86, 105, 128, 132, 157
indirect 69
Speech act Theory 113
Speed of processing 20, 83 – see also processing efficiency
Spelke 171 n.3
Squire 84
Stances 92, 93, 105, 136, 139, 140
Stimulus 61, 67, 72, 84
accidental 80
Stimulus-response connection 20, 61, 77, 79, 158
Striatum 84
Structure
deep 66
lower-level 59, 62, 89
surface 66, 68
Subjective (epistemological and ontological sense) 47
Sweetser 25
Symbol learning 18, 66 – see also prefrontal cortex
Symbolic threshold 22
Synaptic connection (strengthenin) 27, 160, 169
Synaptic changes 170, 172
Synaptic efficacy 169, 177
Synergies 173
Syntacticization 21, 23
Syntax 21, 175
Survival – see also adaptive values

Tagging 103 n.12
Task 69
identification 73, 80, 84, 166
sorting 19
Tendency

innate mental 17
competing 19, 71
inhibition of 101, 147, 160
TNGS 162, 165–167, 176
TNGSx 175
Top-down mode of interpretation 103
Topos 90, 91, 114–116, 139, 140, 153
Traugott 13 n.1, 14, 39, 40, 51

Unconscious 106
Underlying means 15, 26-28, 83–86, 165

Value-category matching 166, 176
Van Essen 85
Verbs
mental 40
speech act 40
Visual perception 75
areas 74 n.10, 85
imagery 75
edges 86
motions 168
Wernicke's area 175

World
external 22, 23, 34, 38, 42, 44–46, 50, 122, 128, 147, 166, 170
inter-subjective 128
subjective 45, 50, 122, 127, 128
uncategorized 167

CONTEMPORARY STUDIES IN DESCRIPTIVE LINGUISTICS

Edited by

DR GRAEME DAVIS, Research Co-ordinator, Arts and Social Sciences, at Northumbria University, UK, and

KARL A. BERNHARDT, Head of the English Language Foundation Programme at the University of Buckingham, UK.

This series provides an outlet for academic monographs which offer a recent and original contribution to linguistics and which are within the descriptive tradition.

While the monographs demonstrate their debt to contemporary linguistic thought, the series does not impose limitations in terms of methodology or genre, and does not support a particular linguistic school. Rather the series welcomes new and innovative research that contributes to furthering the understanding of the description of language.

The topics of the monographs are scholarly and represent the cutting edge for their particular fields, but are also accessible to researchers outside the specific disciplines.

Vol. 1 Mark Garner: Language: An Ecological View.
260 pages, 2004.
ISBN 3-03910-054-8 / US-ISBN 0-8204-6295-0

Vol. 2 T. Nyan: Meanings at the Text Level: A Co-Evolutionary Approach.
194 pages, 2004.
ISBN 3-03910-250-8 / US-ISBN 0-8204-7179-8